D0315688

An Introduction to Video Measurement

An Introduction to Video Measurement

Peter Hodges

Focal Press
An imprint of Butterworth-Heinemann
Linacre House, Jordan Hill, Oxford OX2 8DP
A division of Reed Educational and Professional Publishing Ltd

A member of the Reed Elsevier plc group

OXFORD BOSTON JOHANNESBURG
MELBOURNE NEW DELHI SINGAPORE

First published 1996

British Library Cataloguing in Publication Data
A catalogue record for this book is available from the British Library

ISBN 0 240 51447 5

Library of Congress Cataloguing in Publication Data
A catalogue record for this book is available from the Library of Congress

Typeset by Avocet Typeset, Brill, Aylesbury, Bucks
Printed in Great Britain

CONTENTS

PREFACE

Following publication of my first book, *The Video Camera Operator's Handbook*, I received a number of enquiries about the subject of video measurement. Hence, this much more expansive publication goes on from where the Handbook left off.

It is aimed at new entrants to the industry and those already established but who now feel the need to understand the system better. From the beginning the intention was always to bring everything down to simple basics, not particularly easy with so complex and diverse a subject. But there is little point in overwhelming the reader with unnecessary technical talk when a simple statement or diagram will suffice. So, a few corners may be trimmed ... but the foundation stones are not.

I hope this book will also be a 'good read' for established engineers; we are a profession that can never stop learning and I have learned a lot from writing it.

I want to thank my colleagues in the industry, particularly in the BBC, who have been bounce boards for ideas about presentation as well as theory. Special thanks to Wendy Ross and to Keith Schofield for their help. Manufacturers have also provided assistance; in particular, Leader Instruments (UK) Ltd, and also Steve Nunney of Hamlet Video International Ltd.

Peter Hodges

INTRODUCTION

What do we measure in video, and why?

The answer is the reason behind this book. Video is a high tech business that makes use of engineering to acquire, store, manipulate, send and display pictures. Measurement is essential to engineering. To describe an engineering situation, we must be able to measure and quote all the dimensions. When we talk of pictures, which is the reason for video, it is insufficient to simply say 'it is too dark or too light'. These are photographic terms, perfectly legitimate where the esoterics of picture making are concerned, but not engineering parameters.

Engineering measurement makes sure all the parts fit. In video this means making sure the picture fits the system. Or, turning the argument around, making sure the system can handle the picture. Modern production techniques are always changing, bringing a constant pressure to alter and improve systems. Video engineering has standards set out in volumes published over the years; this is the information used by engineers involved in both design and operations.

We often hear video described as a simple medium; just switch on and use. That's a fine ideal to aim for but there is a vast amount of electronic processing taking place in every programme. Although a lot of this can just be left to get along by itself, some technical supervision must be on hand if errors are not to develop into faults and compromise operations.

The title of the book is an incomplete description for one cannot talk about measurement until it is clear what has to be measured. And this is not possible until the subject is properly understood. Consequently, the first part is entirely devoted to the principles of electric signals and ending with a description of the video system. And even here, colour fails to get a mention until black and white has been thoroughly gone over. Finer points still, have to wait their turn until the specific systems that concern them, are dealt with. Such is the complexity behind the simplicity of video.

The emphasis is on making video engineering a subject of interest and knowledge. Particularly for those who want to know more about the subject but do not possess that foundation of electronics on which to base their study. But by starting right at the beginning and working through camera set-up, studio timing and videotape editing, a pretty broad knowledge of the whole video system should be gained.

Knowledge is, by itself, not enough. Experience is the other vital commodity. No one can teach experience; it has to grow and expand in its own good time. Current working practices are causing experience to leave the industry, either because it's considered too expensive or no longer necessary. Human presence can be engineered out of some areas, of that there is no doubt. But we have yet to reach the point when the technology knows what the eye wants to see, although some would have us think otherwise.

'It never fails' is another myth. No technology can provide that. Testing a piece of kit is no guarantee it won't fail as soon as your back is turned. This is no gloomy prospect, its a fact of life, and one we have lived with for a very long time. This book sets out what has been known and used since electricity came to town. All that has been done is to turn it into a form that anyone with the will to learn can benefit from. To know and understand generates confidence.

The book ends as digital begins. There is a growing inclination to discard 'old video' as if to make way for a new order. This fails to recognise that the new is built on the old and the two are inseparable. One cannot begin to understand how digital video functions until there is a complete understanding of its analogue base. To skip over this by assuming it is dead and gone is a delusion. It still lives on inside its new digital packaging.

CHAPTER 1

CIRCUITS AND SIGNALS

An electric signal is the sending of information by means of an electric circuit. Video is a complex signal that is made up of different elements that describe the picture and how it is constructed. To understand the video signal, we must look at the way a picture is converted into electricity and transmitted. Understanding this requires some knowledge of electrical and electronic theory. So let us start right at the beginning with the simplest of electrical signals.

Signals and transmission

Take a simple circuit; that of a lamp powered by a battery. Figure 1.1 shows the battery and a lamp. A switch is also included. This is a two-wire circuit; one leaving the battery and one returning to it, at the lamp, there is one arriving and one leaving. The switch is placed in one of these wires – it does not matter which. Figure 1.1 shows both sketch form and schematic form. It is convention to use schematic diagrams with international symbols to describe a circuit. We will follow this practice with explanation as required.

Sending a signal

First of all, what is a signal? When a switch is turned on and off, it sends an on/off current down the wire. Where an electric current in a wire is varied into a form that replicates audio or video (or, any other sort of information) it is called an electric signal. The form chosen will depend on the system or electric circuit to be used and the information to be sent. The means to detect the information in the signal and reproduce it will be required at the end of the wire – the **destination**.

There are many types of circuit, depending on what signal system is to be used and the kind of information it must carry. All may be considered as having evolved from the battery, switch and lamp circuit. Sending, or transmitting, the signal may originate with a two-wire circuit, and end as a two-wire circuit, but in between, there may be other forms of transmission, e.g. a wireless circuit, as well as the switch.

Injecting a signal into a length of wire is like launching a missile. It can be aimed with proper regard for its size and shape, how far it has to travel and through what medium. Or, it can be hurled without thought for any of these. Our lamp and battery circuit would be in the latter category if

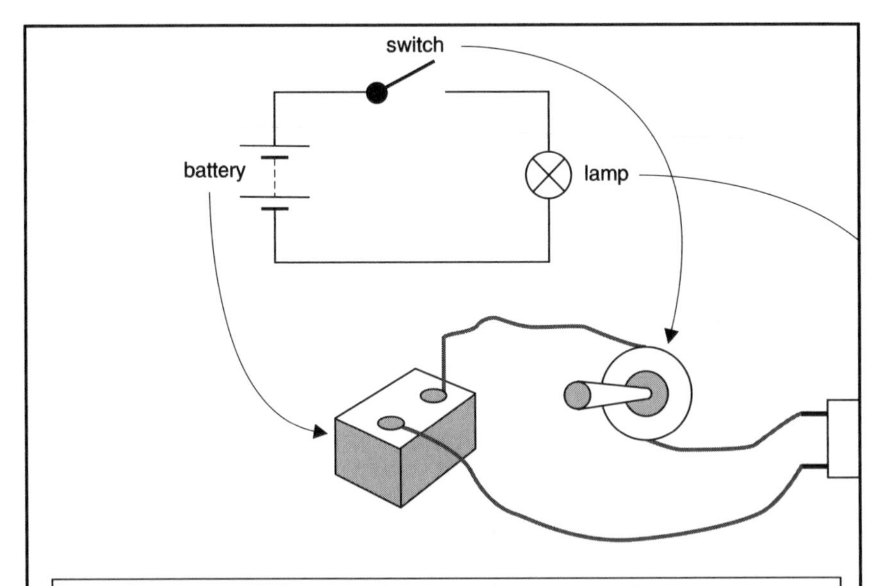

Closing the switch places the battery voltage across the lamp, driving a current through it. The lamp filament heats up and radiates light. This is a straightforward transfer of energy from electricity in the battery to heat and light. When used as a **signalling system**, the information transmitted is at its most basic for there are only two conditions; On or Off.

Figure 1.1a When the switch is closed the lamp lights

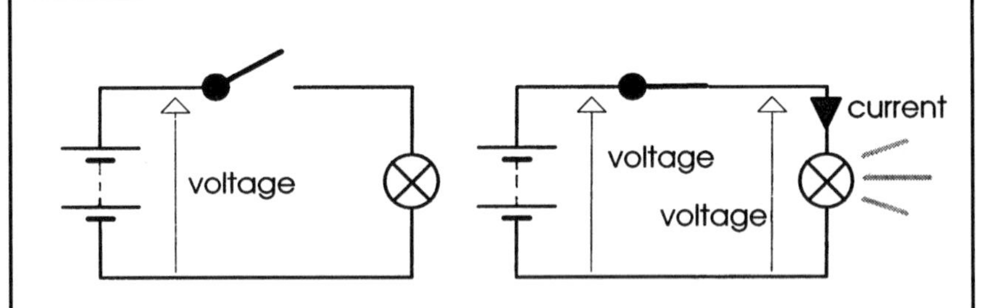

Measurement in this circuit is simple: we can measure the voltage and the current and from these, know the whole of this circuit's values or parameters. Battery voltage is always present but closing the switch effectively places the voltage across the lamp. The lamp forms a **load** on the battery voltage that causes a current to flow in the circuit. Battery energy is now transferred to the lamp, appearing as heat and light.

Figure 1.1b Two-wire electric circuit conditions

just made up of bits of wire and insulating tape. Although able to signal, on or off, its usefulness is obviously limited.

Even so, the comprehensive world-wide system of cable telegraphy is based on this simple circuit. Using a battery and switch to send a signal or current along the cable or wire and a sensitive indicator to replace the inefficient lamp, messages in simple code are sent under the sea and over land. While the 'missile' here may be thought of as crude and the path it passes along little better, cables have now been in use for over 150 years. So proving it to be an exceptionally reliable way of communication. Couple this to its inherent security and resistance to interference and it is easy to see how valuable the cable telegraph was to become in times of international stress. A point to reflect on with the rapid growth of hi-tech eavesdropping on wireless, satellite and software.

The complexity of today's signals as used in sound and video require greater sophistication in their handling and transmission than yesterday's cable telegraph. In practice, the circuit is usually a *fait accompli*, often provided by a contractor. The actual **line** (as the cable is usually called) can be taken as conforming to a standard. Alternatively, transmission without wire into free space is also established and well documented. But the longer the line, the more probable the chance of error or fault. Particularly where the means of transmission changes from one system to another.

Comparing the circuit of battery, switch and lamp to a telegraph cable would be perfectly reasonable but the presence of a long interconnecting cable has to be considered. No cable or piece of wire is ever perfect. For short lengths, the metre or so of wire that makes up the circuit in Figure 1.1a, there is negligible effect on its operation. So simple a concept, however, starts to break down as the distance between the signal source and destination increases. In such a case, the connecting wire, whilst carrying the signal current, absorbs significant power from it.

Figure 1.2 shows the schematic diagram and describes the operation of sending a simple 'OFF' to 'ON' transition of the signal. Note that there is a time difference from when the switch closes, sending the signal into the cable, to its appearance at the destination. This is due to the finite transmission time taken by an electric current in a circuit. Although very fast, approaching the velocity of light, in fact, the effect is significant in electrical terms over long distances.

When our battery voltage is applied at the sending end of the cable, the first thing that happens is the cable 'starts to fill up with electricity', a statement that simplifies the action but serves to illustrate the mechanism. The cable can, therefore, be said to have a capacity which must be filled before a signal can emerge at the far end. One can compare this to filling a pipe with fluid; quite a lot of fluid will go into a long pipe before any reaches the far end.

Like all simple analogies, this one can be subject to misinterpretation, so avoid taking the idea too literally. It does, though, offer an alternative way to understanding the concept of **signal transmission** over long distances in basic terms.

Source and destination

In an ideal world we would be able to say that the signal appearing at its destination will be identical to that sent out from the source. From the discussion so far, we can see that this is not possible. It is however, an ideal to be aimed at and, as long as the degree of degradation or attenuation is known and is acceptable, then a practical and predictable signal communication system can be achieved.

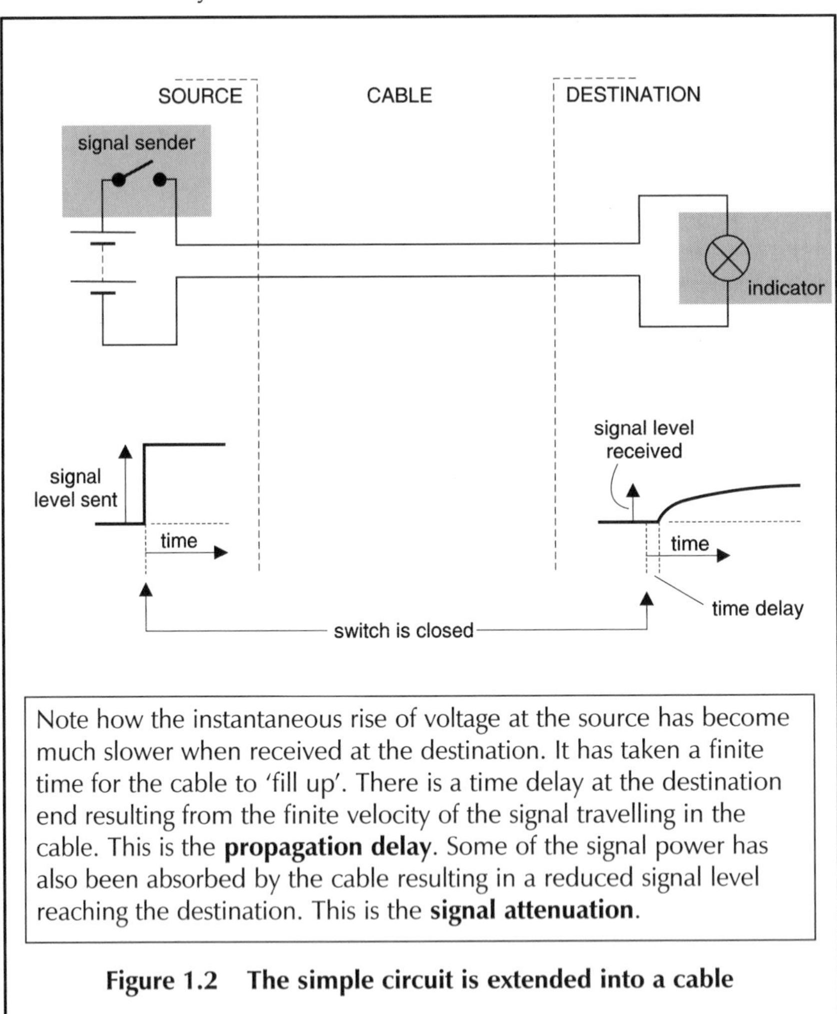

Figure 1.2 The simple circuit is extended into a cable

But what do we aim for in trying to achieve an ideal system? This question is the bottom line of all signal transmission so let us list a few of the most important circuit ideals:

1. The conditions, or characteristics, at the source and destination are known.
2. The circuit cable and associated equipment are compatible.
3. The circuit characteristics are known.
4. The signal and circuit are compatible.
5. The signal is able to carry the information required.
6. The signal will be immune from interference
7. The circuit will be reliable.
8. The information will be secure.

Items 1 to 5 embrace the technical requirements of the circuit design. Ideals 6 to 8 are largely dependent on the kind of circuit used and here, reference to 'circuit' may in practice, go beyond the two-wire circuit considered so far. It is quite usual over long distances to have combinations of cable, wireless or satellite-based systems. But the first ideal states that the source and destination characteristics should be known. If we also know item 3, which sets out the characteristics of the circuit, we can predict how the circuit will react to, and alter, the signal.

Receiving a signal

Figure 1.2 shows the signal distortion that occurs over a length of cable. It is assumed that the cable design is optimised for the job it has to do. The two significant parameters of the cable are:

1. **Cable resistivity**: this causes signal power loss.
2. **Cable capacity**: how much electric current flows into it before the signal can appear at the destination.

In addition, Figure 1.2 also shows the propagation delay – how long the signal takes to appear at the destination.

Cables may be used over kilometres or just centimetres. The length will inevitably have considerable effect on the signal and its distortion. All cables, whatever the length, will affect the signal to some degree, but by careful design the effect can be minimised and made predictable. Cable design influences, in particular, how fast the signal can send information, as well as how far. This is the cable **characteristic**. An integral part of a cable characteristic is to state how the cable is loaded, or how the cable is **terminated**.

Terminating the cable

In the basic battery/lamp circuit, power is transferred from the battery to the lamp when the switch closes. How all this works was described years ago by Georg Ohm and his law is still the most universal in electric and electronic theory. He stated a circuit has three elements:

1. Voltage. The battery has a defined voltage.
2. Current. The voltage drives an electric current into the load.
3. Resistance. The load offers resistance to the current and heats up in the process.

These three parameters are related by Ohm's Law:

Voltage = current x resistance

All electric and electronic circuits possess these three elements and conform to this law.

When current passes through the load it may heat up because **power** is dissipated in it. A signal is received as power; the power to operate a lamp or indicator. In more advanced circuits, this is the power to produce sound or pictures, usually with the aid of additional apparatus. Power cannot be sent unless there is a load, or resistance, to accept it.

Power is dependent on the voltage applied and the current flowing in the load – the product of the voltage and the current:

Power = voltage x current

The signal voltage may be present but no signal current will flow until a load is present. The destination load terminates the cable to complete the circuit. In the case of a long cable there will be a finite time before the load is 'seen' by the signal. The sending of power is, therefore, not possible at that instant the switch closes. Figure 1.3 shows this effect.

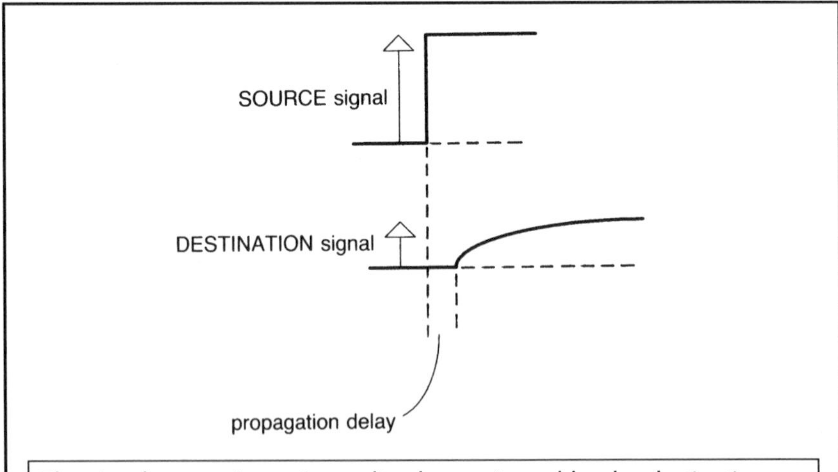

The signal source is not immediately terminated by the destination load because of the propagation delay. For the signal to be launched properly, the design characteristic of the cable must provide the initial load so allowing the signal to enter the cable.

Figure 1.3 Propagation delay

If the cable characteristic and terminating load are correctly chosen, the cable itself will provide the correct load for the signal at the source. The cable, source and destination are now said to be **matched** allowing the signal to be sent with minimum loss and distortion. In practise the cable has to be *correctly terminated at both ends*, meaning that source and destination are required to match the characteristic of the cable.

The subject of cable termination will arise in later chapters for it is a very significant part of video transmission.

Creating a standard

Where there is a common cable and termination characteristic as described above, there is the basis of a circuit or transmission standard. In creating a transmission standard, we put in place the means to send and receive signals to and from any point that conforms to that standard. When proved and accepted as a good standard, it can be promoted for the use of all. A standard enables interconnection, not only for long distance transmission, but simply to connect one piece of equipment to another. Wherever the standard exists there is the means to interconnect.

The value of this is obvious; system design is simplified, it is flexible and, therefore, very cost effective.

The next important consideration is the ability to check the system. The standard specifies all the parameters of the signal and the transmission circuit, including the cable and its termination. The standard is a fine ideal and quite crucial to reliable and repeatable signal transmission. There are, however, many pitfalls for our signal on its route and we have to ensure that the signal is generated accurately and conforms to the standard in all respects. Signal measurement is a very significant part of this. This is required at the source, where the signal is generated and, on arrival, at its destination. The signal may be subject to further checks *en route*, depending on the how far it has to travel and by what method.

The integrity of the standard has to be maintained if it is to be successful. It therefore, depends on all users to ensure that it is operated properly and is not allowed to fall into disrepute. Signal measurement is a core element of this. Measurement may be an ordered discipline, carried out every time a circuit is used. Or it may be less formal, no more than confirmation from the destination that the signal is being received.

Which procedures are most appropriate depends on:

1. The length of the transmission path.
2. What route and how many stages in the circuit between source and destination.
3. How likely the signal is to be distorted or damaged.
4. Whether or not the circuit is one of proven reliability or not.
5. The importance of the information sent.

The first point is easy to evaluate; a short length of cable will not require the attention devoted to an international circuit. Point 2 is related to the first, the greater the complexity, the more attention will be required. The third is often dependent on the first two and a measurable quantity. Point 4 is less predictable and here, experience becomes significant. The final point can only be evaluated by the user, not the operator. The distinction here is important.

The standard allows for the inevitable errors in our practical systems to be identified. The signal can be generated precisely and measured. Upon its arrival a similar check can be made to verify its condition and action taken if below the standard. Here lies the fundamental reason to measure; to know that what is sent is correct and agree that what is received is also correct.

Cable and wireless circuits

Figure 1.4 shows how a typical modern circuit has developed from the basic two-wire cable in Figure 1.2. Both source and destination conform to a standard and neither sender nor recipient need to know the characteristics of the rest of the system. Sender and transmitter will be **matched** by the standard, the signal delivered by the sender is precisely what the wireless transmitter requires to convert into a signal suitable for the antenna. At the destination, exactly the same applies, the receiver takes the signal from its antenna and converts into a form the indicator requires.

Using the principle of standard interconnections, the system now lends itself to analysis by the test, or measurement, set-up. The set-up can be used to test the accuracy of sender and indicator. Simply connect them together at (A) and (B). Although these may in practice be some undefined distance apart, it is reasonable to have an identical receiver at the source for just that purpose; testing and measurement. Of course, such a method demands the two receivers are identical, at least as regards point (B). But that is what the standard does. It defines the characteristics of the equipment interconnecting points.

Figure 1.5 goes one step further; the indicator is directly connected to the sender. This step is significant for it makes apparent the need to adapt the standard signal to whatever method of transmission we are using. Both sender and indicator conform to the standard; they are interconnectable.

Cables and wireless antennae need specific equipment to operate them. A cable requires a **line driver**, whilst a **wireless transmitter** will drive an antenna. The destination will require a **line receiver** or a **wireless receiver**. And our simple cable circuit in Figure 1.2 has developed into a more universal one.

A standard becomes established when users recognise it as a successful one, encouraging its use and expansion. Eventually all equipment and systems fall into line and base their interconnections on that standard. A whole range of options may now be made available for different operations.

As has already been shown in Figure 1.4, a complete transmission system may be a combination of techniques. We have seen that the presence of a circuit at the source does not necessarily mean a cable travelling **all**

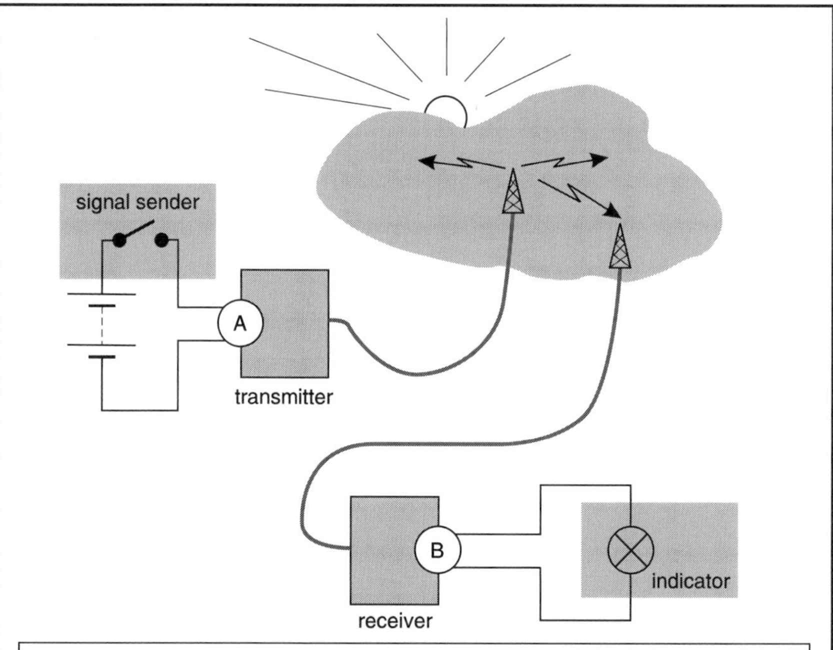

Here is a typical transmission circuit, it may be intercontinental or simply from inside a sports stadium to the truck outside. But note how the sender and indicator no longer 'look' into a long length of cable. At point (A) there is now a transmitter. This effectively separates the 'outside world' from the sender. Likewise, at Point (B) there is a receiver isolating the indicator. Both these devices, of course, serve the primary function of transmitting and receiving, via their antennae. The characteristics of terminals (A) and (B) can now conform to a standard, make them have the same characteristic and we are back to our original two-wire circuit as regards the circuit as 'seen' by both sender and indicator.

Figure 1.4 The wireless circuit

the way to the destination. It is only the standard at source and destination that must be the same if signal accuracy is to be guaranteed. To send anywhere, whether to the other side of the globe, or just to the room next door, can now be undertaken with the confidence necessary in modern communications.

Increasing complexity

So far we have only considered a signal based on a battery and a switch as the sender and the indicator, which may be a lamp or sounder. For more complex signals, such as audio and video, the principles of the circuit and its characteristic still hold true. The additional complexity of the signal place additional demands on its precision of generation and its handling by the circuit, which must therefore, improve and be maintained to a higher level. For the basic principle of a standard is that, once established, it must always remain true.

Such is the complex nature of audio and video that errors may be many, significant or otherwise. To ascertain the significance of these can only be done with proper measurement and understanding.

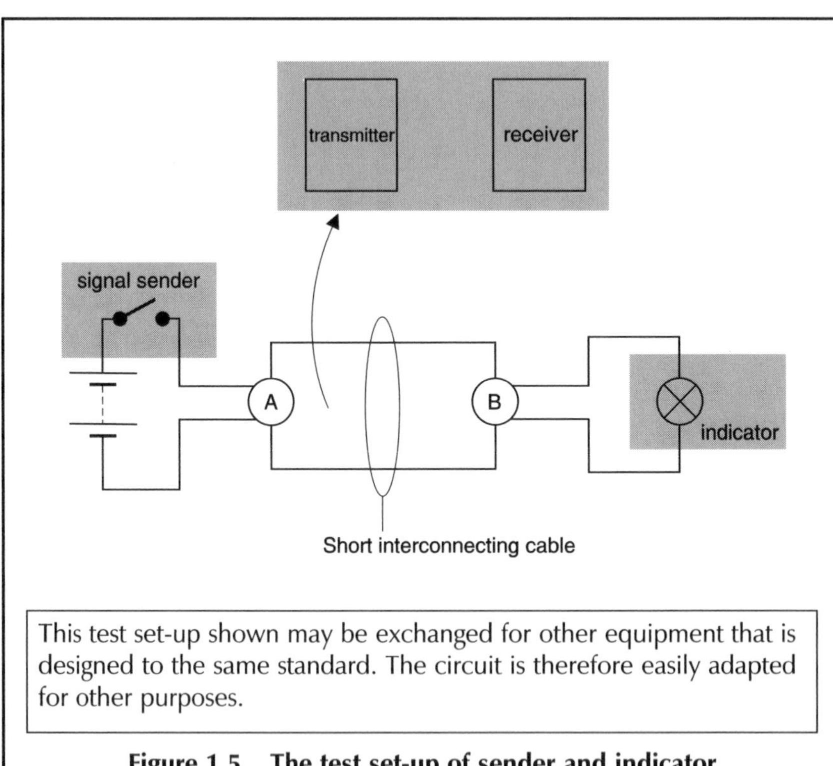

Figure 1.5 The test set-up of sender and indicator

CHAPTER 2

TIME AND COMPLEX SIGNALS

The complex signal

The more complex the signal, the more stringent the requirements of transmission. Generally speaking as the rate of information increases then so do the demands on the circuit. How do we define 'information rate'? Return again to cable telegraphy based on the battery and lamp circuit. Here, the rate is the operator's speed in converting the information into a code of on and off signals, followed by the reader's speed in decoding again. The code is rather similar to Morse Code and is much slower than speech, many, many times slower. Speech signal fluctuations far exceed the speed of the battery and switch system.

Speeding up the transmission of cable telegraphy is quite possible. Firstly, replace the human operator with a faster automatic system of switching. Then, redesign the circuit to function at the higher rate and, finally, install a high speed reader at the destination. This may all seem very logical, but above all, the desire was to send speech and so the Bell telephone system caught the imagination. High speed data transmission, for that is what an automatic rapid on/off system really is, was not introduced until afterwards.

Speech was not easy to send over long distances. It was more prone to distortion and interference. But so attractive was the idea that a very great deal of effort went into developing the means to make it possible. Audio, or speech, transmission systems became known universally as 'telecommunications'.

Video is a relative newcomer. It was able to build on all the audio research, inheriting many features of the audio telecommunications. Not least was the nomenclature, or how the various components came to be described. We will, therefore, concentrate on audio, for the examination of audio signals and how they are carried, hold many clues to dealing with the complexity of video.

Figure 2.1 shows a typical sound wave. It commences with a sharp increase in air pressure followed by a slower decay. The shape, or **waveform**, is of a noise that could be described as a crashing or breaking sound. Note how irregular the shape is, typical of the sound created by a mechanical action. Every such event will have its own distinctive form, sometimes called a **sound signature**, which is never exactly repeated however many times the event may be repeated.

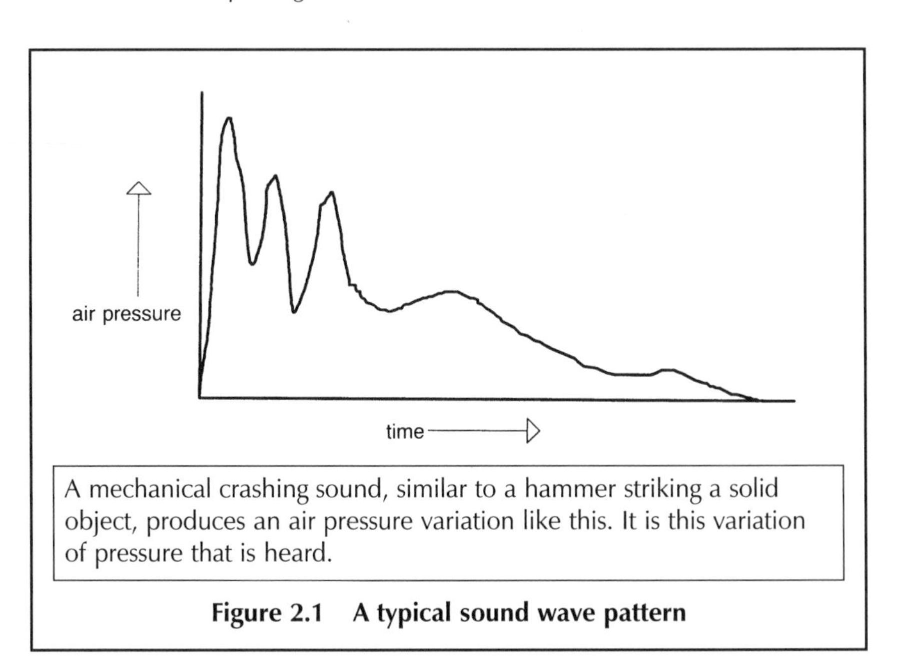

A mechanical crashing sound, similar to a hammer striking a solid object, produces an air pressure variation like this. It is this variation of pressure that is heard.

Figure 2.1 A typical sound wave pattern

Whilst the waveform in Figure 2.1 would be heard as a crash, the waveform in Figure 2.2 will be heard as a steady pure tone. Where the first example is an event with start and finish, the second is continuous. Figure 2.2 describes a **sine wave** and represents the purest form of sound. The simple motion of a pendulum is the classic mechanical sine wave, also known as **simple harmonic motion**. Sine waves are not confined to mechanical systems, they are fundamental to electrical circuitry as well.

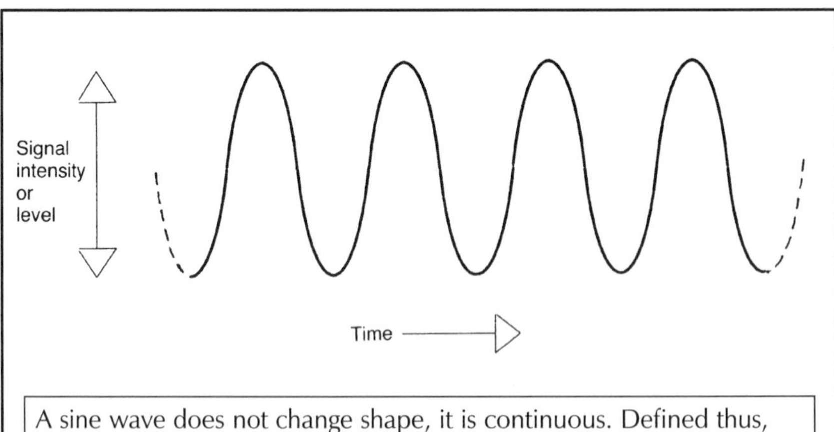

A sine wave does not change shape, it is continuous. Defined thus, without change, it would theoretically continue forever. Note that the vertical scale of 'signal intensity' has a double-ended arrow indicating a variation about a mean value.

Figure 2.2 A sine waveform of pure tone

Figure 2.2 is of a sine wave signal. This is a graph that shows signal level and time just as in Figure 2.1, but this wave is represented as continuous, no start or finish is specified. Neither axis has a reference scale, nor are there units but the arrows indicate the direction in which they increase in value. Before we can make any measurements of these signals, the units of time and signal level must be set out with reference scales.

All this may seem academic, but where the more complex waveforms associated with video are considered, these become fundamental to thorough understanding.

Three sine waves are shown in Figure 2.3. They are all related, their start times coincide at the minimal level at point (A) and all return again to the same level at (B). From (B) the cycle repeats. They differ in only one respect, their **frequency**. Frequency describes how many times per second the signal repeats itself, or completes one cycle. The lowest frequency wave completes one cycle from (A) to (B). Another completes two cycles and the highest frequency wave completes four cycles, all in the same time.

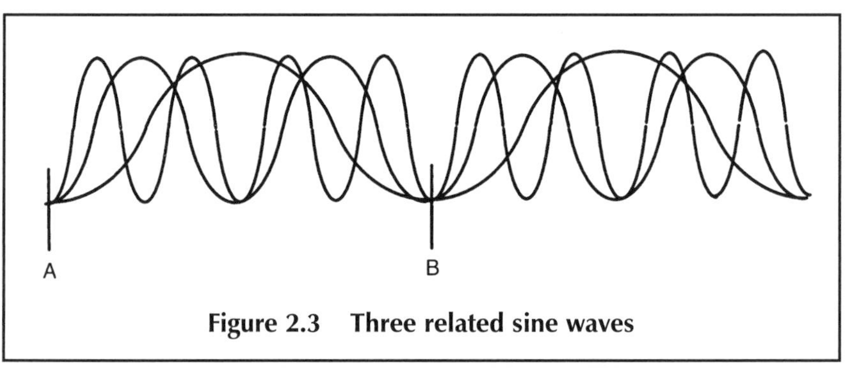

Figure 2.3 Three related sine waves

Frequency, is specified in cycles per second. The term 'cycles per second' is now superseded by 'Hertz', abbreviated to 'Hz', after the German physicist who first identified radio waves. One Hertz is one cycle per second, usually abbreviated to '1 Hz'.

The representation of three distinct waves, as in Figure 2.3 is a theoretical situation. Sound pressure at any particular point in space and at any particular instant in time, can have only **one** value. Therefore, for all three to exist in the same system, they must add mathematically to produce a more complex wave. The resulting waveform is not sinusoidal and appears as in Figure 2.4.

One may see a similarity between the waveform in Figure 2.1 and that in Figure 2.4. There is irregularity, neither is symmetrical about the horizontal, or 'time', axis. In fact, the waveform in Figure 1.2 may be constructed from individual sine waves. However, to build, or synthesise, such a signal requires a very complex combination of sine waves. The theory quite clearly states that this is possible and, indeed signal synthesisers are a common feature of audio practice. Should there be a desire on the part of the reader to study the theory further, there are numerous

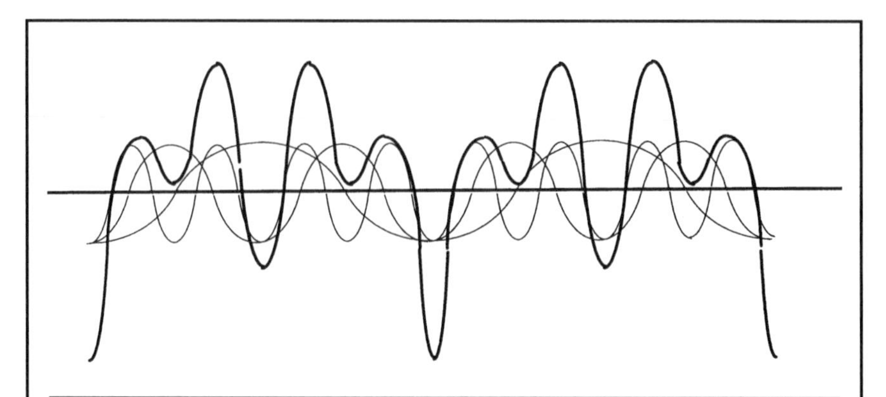

Adding together sine waves produces a sine-based resultant whose variations either side of the mean are not always symmetrical.

Figure 2.4. The three sine waves in Figure 2.3 added together

books available to do just that. Suffice to say that any sound may be broken down into its constituent sine waves.

So far, we have only considered variations of signal intensity as a sound pressure variation, and only in *one direction*. The sudden pressure increase as illustrated in Figure 2.1, is followed by a decay. If we were to continue the observation, we would see the pressure reverse for, like a

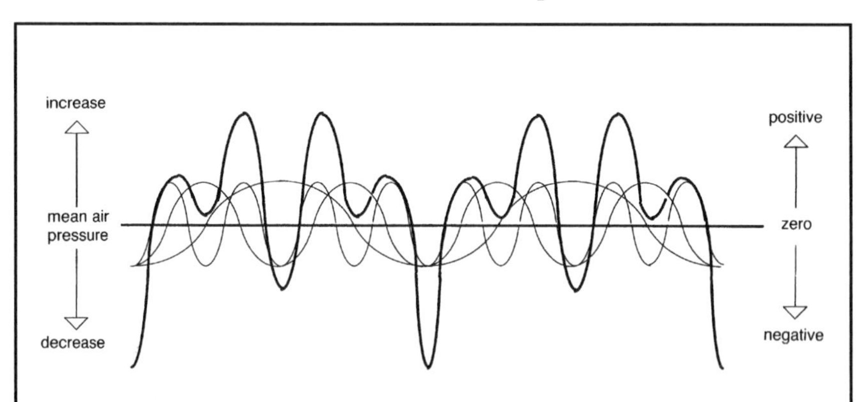

Two scales are shown for the vertical axis. The left-hand one is air pressure variation about a mean value. On the right-hand side the mean value has been subtracted. The pressure variations may now be considered as positive and negative of the horizontal axis. Note that the 'time' axis is not shown, it is often assumed but not stated, unless the time scale is relevant.

Figure 2.5 Defining the values of the axes

wave on water, a flow in one direction will slow down and eventually reverse. This is an oscillating system. Oscillating mechanical and electrical systems will swing from one side to the other as long as energy remains in them to continue. Eventually, the energy will be used up and the system will become stable once more at the mean air pressure.

Figure 2.5 shows a vertical axis scale has been added with a reference line above which the pressure rises, and vice versa. This translates into positive and negative values of the signal. This is accepted convention, positive above and negative below the reference line. The reference line now becomes, by definition, zero. In sound pressure terms, the line represents normal air pressure about which the signal swings, but it is usually more convenient to ignore this offset, referring to the signal as swinging positive and negative about this value.

Converting to an electric signal

On converting the acoustic signal into its electrical equivalent produces a varying voltage of similar shape. Figure 2.6 has no units specified for amplitude and as (from Ohms Law) voltage and current are proportional, either may be assumed. Voltage has become the established value of signal transmission and time in seconds, or fractions of a second.

Remember that when using voltage as the unit of signal level, we must never lose sight of the fact that it is the *signal current that carries the infor-*

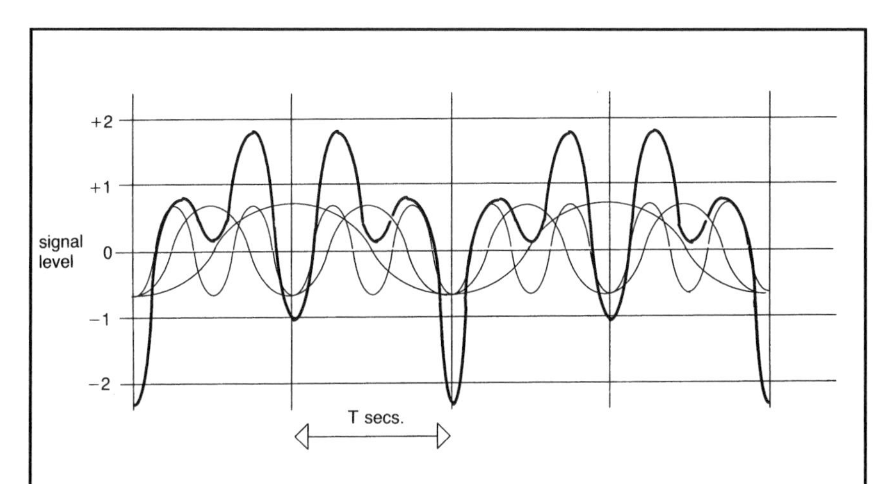

Placing a scale of values against the waveform enables it to be measured. Here, signal level is scaled from +2 to –2, called 'amplitude'. T is a value in seconds between the horizontal spaces, as time is continuous, no start or finish is stated.

Figure 2.6 Defining the values of the axes

mation. Either a steady direct current, as in telegraphy, or the fluctuating speech signal, both are subject to the relationship of voltage, current and load resistance.

Figure 2.6 is a complex signal. If the values of its component frequencies were stated as being between 20 cycles per second and 20 000 cycles per second, it could be an audio signal. This is the frequency band of human hearing. Therefore, the **frequency range** of high quality audio is often stated as being 20 Hz to 20 000 Hz (other values are sometimes used but such differences need not concern us here). As far as the ear is concerned, the sound in Figure 2.1 can be represented by a combination of sine waves whose frequencies will lie between 20 and 20 000 Hz.

When audio is converted into an electrical signal it may be described as **alternating current**. A battery can only deliver a signal that is a **direct current**, i.e. it always flows in one direction as determined by its chemistry. Thus it has a **polarity** indicated as positive and negative. Direct current has two units; current expressed in **amps**, and **volts** (the force driving the current). Alternating current also uses the same units of amps and volts, but, as we have seen, there is the added parameter of **frequency** to consider. Frequency may be stated as the rate at which the polarity alternates.

Therefore a signal based on an alternating current has three parameters that may be used as carriers of information. In practice the voltage and current are interrelated and are not, therefore, available as independent information carriers and it is usual to consider voltage only.

With voltage we can state a scale of level from, say, zero to 1 volt. This is quite arbitrary; there is no reason why designers cannot use any other set of values to suit their own individual requirements. Inside equipment, on the other side of the connectors, any value of voltage may be present. The choice is entirely the designer's. It is only where the signal enters and re-emerges again that we need be interested in actual values of signal level and how it must comply with the standard.

The concept of time

The original starting point of time is now a very long while ago and so we have to establish our own points of reference from which events may be measured. For instance, we can make a reference at which a particular event in the signal takes place, this may be stated as: $t = 0$ secs. (The standard unit of time is the **second**.)

Figure 2.7 has a single horizontal axis with a scale in seconds. Note that, because time is continuous, the scale commences before the first event, passes through zero and runs on beyond the last event. The reference from which the measurement is made is where t is zero. The two significant events are the door opening and the person speaking. The clock must be able to identify both these events so it is necessary to present them to the clock in a form it can recognise so that it may be stopped and started at these points.

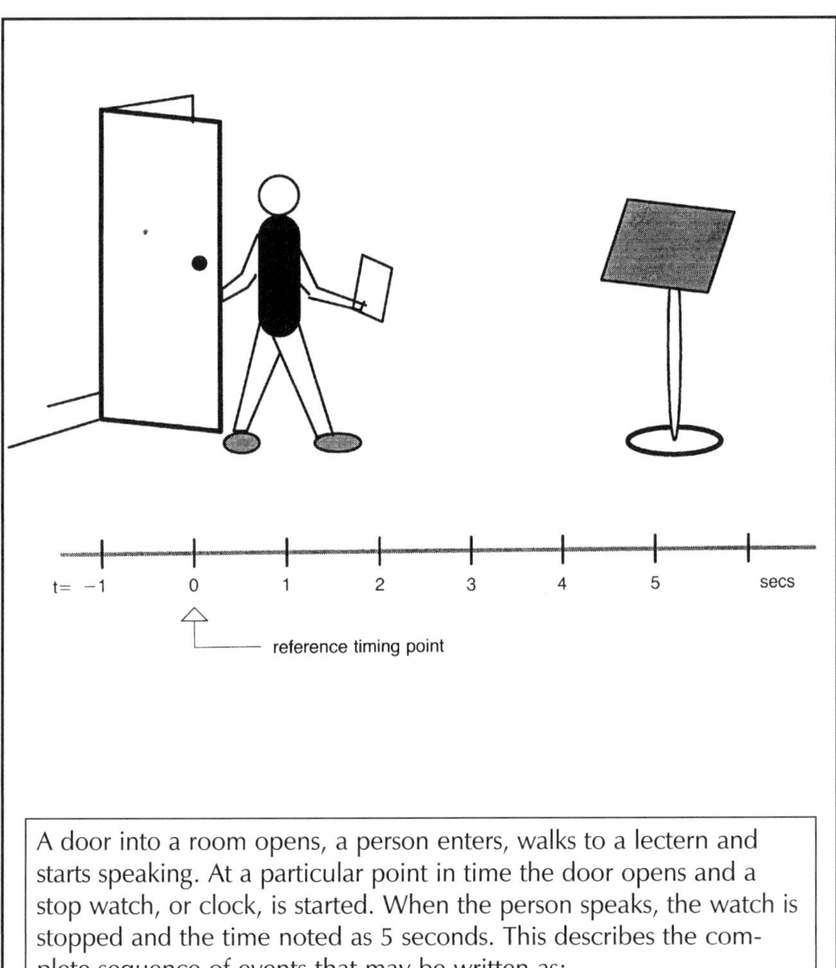

A door into a room opens, a person enters, walks to a lectern and starts speaking. At a particular point in time the door opens and a stop watch, or clock, is started. When the person speaks, the watch is stopped and the time noted as 5 seconds. This describes the complete sequence of events that may be written as:

START:	At t = 0 seconds	Door opens
	t = +1 second	Door closes
From t = +1 second to t = +4 seconds		Person walking to lectern
	t = +4 seconds	Person reaches lectern
FINISH:	t = +5 seconds	Person speaks

Figure 2.7 A timed sequence

But also note how our simple picture has more in it than at first thought necessary. The time-scale only has to cover five seconds, yet the one shown exceeds this. It's like telling a story: we need a little background to understand the whole. Where there is a sequence it is often useful to see a little extra at both ends to satisfy ourselves that we see sufficient to know the full story and nothing of significance exists just outside our immediate field. The importance of this will become apparent as we proceed.

Figure 2.8 is a sine wave with a time scale placed against it, the same scale as that in Figure 2.7, showing four cycles. These may be described as events as before but because they are repetitive the term 'cycle' is more appropriate. Of the four cycles shown, three are within the timing sequence as specified by the scale (the fourth runs beyond the scale). From this we are able to determine the frequency.

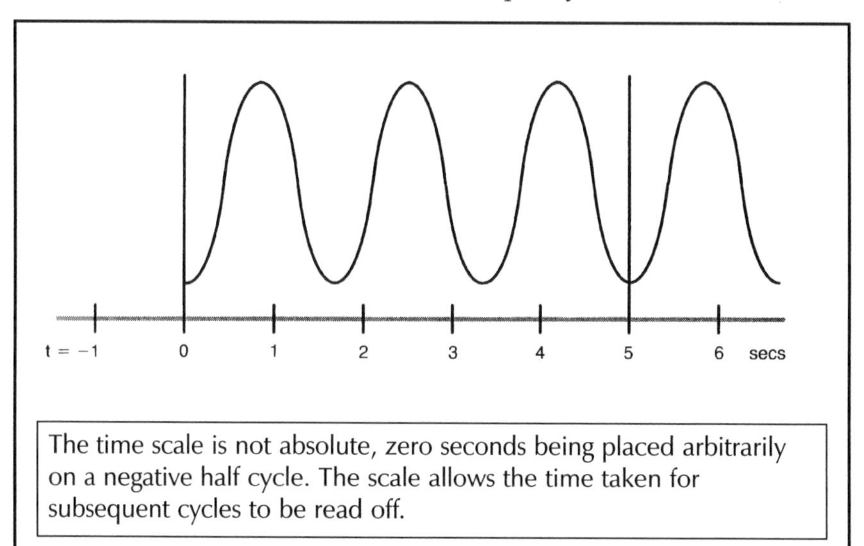

The time scale is not absolute, zero seconds being placed arbitrarily on a negative half cycle. The scale allows the time taken for subsequent cycles to be read off.

Figure 2.8 Adding a time scale to a sine wave

Frequency, f, is defined as the number of cycles per second, so:

$$f = \frac{3 \text{ (no. of cycles)}}{5 \text{ (seconds)}} = 0.6 \text{ cycles/second} = 0.6 \text{ Hz}$$

Now return to the basic battery and lamp circuit that opened this chapter.

Compare Figure 2.9 to that in Figure 2.1. There is a similarity, look at how the Graph (C) is similar with the rapid increase in level at t seconds. The same analysis may also be made of Figure 2.9 as was made of Figure 2.1 regarding its relationship to harmonic motion and the sine wave. Exactly the same analysis may be applied to Graph (A). The battery produces a steady voltage, there is no variation, cyclic or otherwise. The battery voltage is said therefore to have a frequency equal to zero.

This is a very significant point to make; the lowest limit of frequency has been determined, that of a direct current. The upper limit does not exist in practical terms for it is only reached at **infinity**. In our ideal circuit the voltage rise is instaneous when the switch closes.

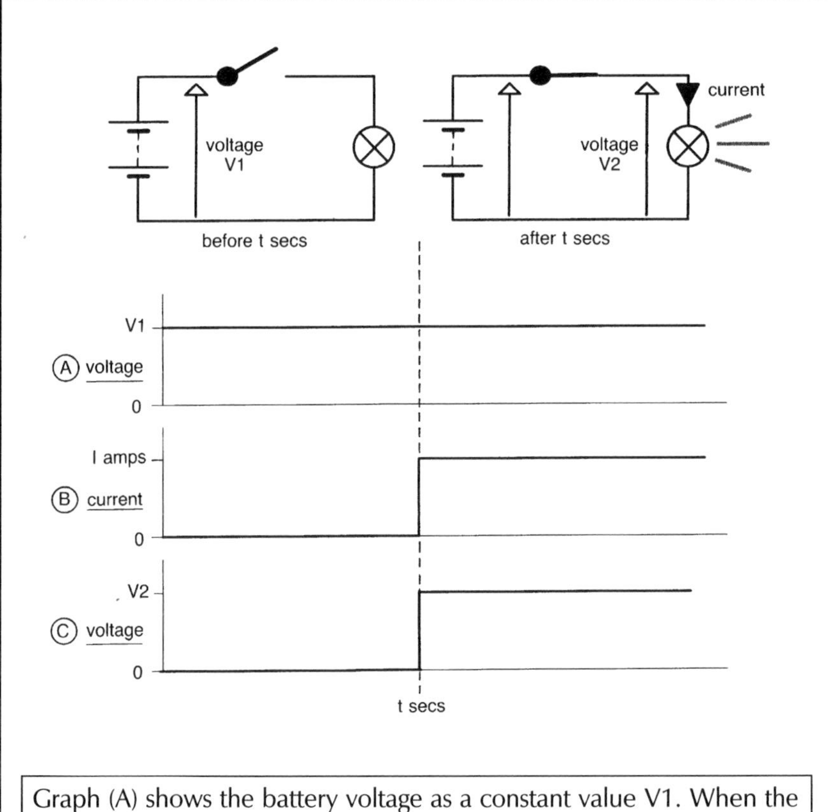

Graph (A) shows the battery voltage as a constant value V1. When the switch closes at time t seconds, V1 appears across the lamp. Graph (B) shows the current increasing from zero to that drawn by the lamp after t seconds. Graph (C) is the voltage V2 across the lamp. Note that after t seconds, V2 increases to equal V1.

Figure 2.9 Timed sequence of a switched lamp and battery circuit

Bandwidth

Of all the various signals considered so far, that of direct current (DC.) and the single sine wave, or pure tone, are special cases for each consists of only one frequency. All others are combinations of sine waves of various frequencies and levels. We have made little attempt to investigate the respective values of these constituent elements, apart from those derived from sound. Audio has a frequency range of 20 Hz to 20 000 Hz, these are said to be the limits of the audio band of frequencies, known as the **audio bandwidth**.

Audio bandwidth = 20 000 – 20 Hz = 19 980 Hz

As frequencies of a very wide distribution are being considered, now is a good time to look at abbreviations. Cycles per second have been given the term 'Hertz' which has already been abbreviated to 'Hz'. To this may be added a whole series of prefixes to simplify the expression of large numbers. Frequencies over 1000 Hz are usually abbreviated to kilohertz (kHz). As 19.98 kHz is so close to 20 kHz, the audio bandwidth is usually approximated to 20 kHz.

Intelligible speech can be transmitted using a considerably smaller bandwidth than 20 kHz. 'Telephone quality', as it is called, typically has a band limits of 300 Hz and 3400 Hz, which is a bandwidth of 3.1 kHz.

At this point, it is worth looking at sending more than one signal down a cable at the same time. The telephone system established a very simple principle, shown in Figure 2.10. The two microphone signals travel round the circuit, carried on the current set up by the battery. The two signals together occupy a bandwidth of 3.1 kHz.

The two-wire telephone works well for two users. More users can be installed into the circuit as well, they can all enjoy a communal conversation. Where secrecy is required, then this simple circuit falls down for the signals are not isolated from each other. Overhearing or interference of one signal by another can only be eliminated when each signal has its own circuit, or **channel**.

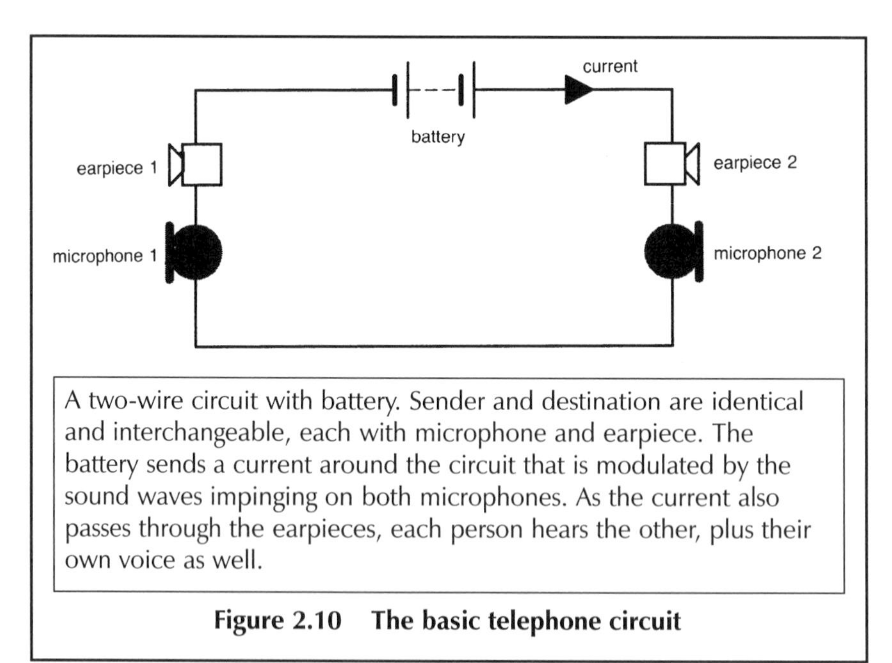

A two-wire circuit with battery. Sender and destination are identical and interchangeable, each with microphone and earpiece. The battery sends a current around the circuit that is modulated by the sound waves impinging on both microphones. As the current also passes through the earpieces, each person hears the other, plus their own voice as well.

Figure 2.10 The basic telephone circuit

Providing more channels means increasing the rate that information is sent. For instance, two independent channels of speech will require two lots of 3.1 kHz each when encoded for sending over a single circuit. Such a circuit would require a total bandwidth of 6.2 kHz This figure is derived from a hypothetical case but gives an indication of how information and bandwidth become interlinked.

So there is a choice; if we want to send more information in a given time, we either increase the number of circuits or the bandwidth.

The bandwidth figures, as stated above, do not, in themselves, state the actual frequency limits of the circuit concerned. For instance, one or more audio channels, each with a bandwidth of 3.1 kHz, could be placed anywhere in the whole frequency spectrum. Figure 2.11 shows six channels, each with 3 kHz of bandwidth coded together onto a **carrier signal** with its own frequency of 1 MHz (one million cycles per second).

The carrier used in Figure 2.11 would be more complex than a single channel of 3 kHz. Its bandwidth points to this fact, it is six times that of one channel. But 18 kHz occupies only 1.8% of the spectrum at the operating frequency of 1 MHz. This shows how, by using higher and higher frequency carriers, more space becomes available to individual signal channels. The principle of using one circuit for more than one signal is known as **multiplexing**.

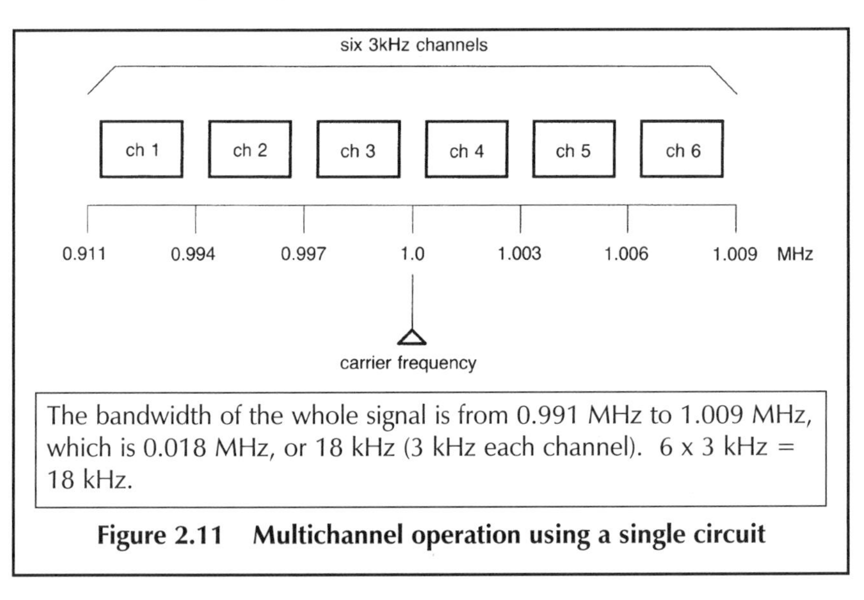

Figure 2.11 Multichannel operation using a single circuit

The principles of bandwidth are most important to understanding how audio and video are sent from one point to another. The effects are far reaching. The voltage step produced in the circuit of Figure 2.9 illustrates an unrealisable situation. As drawn, the signal has a voltage step that rises instantly, implying, from sine wave basis of all waveforms, that this would require an infinite band of frequencies. Such a condition is not possible, there will always be a limit to the bandwidth available.

To observe the true effect we must magnify the part of the waveform at the t seconds point.

Two things can be gathered from Figure 2.12. Firstly, the actual rise takes place *after* t seconds and, secondly, the shape of the rise. Note how there appears to be a part of a sine waveform present. Figure 2.13 shows how this comes about.

From all this we can decide the bandwidth of the signal. Before and after t seconds the waveform is steady (for this example, we can ignore the OFF state before t seconds), which is DC, zero frequency. At t seconds the frequency rises to f Hz. This then is the bandwidth of the waveform; f Hz.

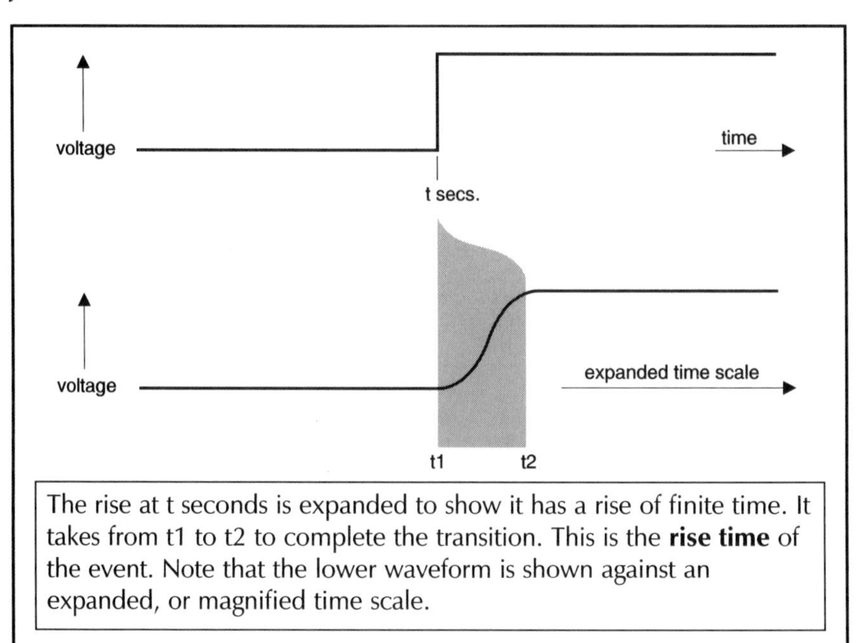

The rise at t seconds is expanded to show it has a rise of finite time. It takes from t1 to t2 to complete the transition. This is the **rise time** of the event. Note that the lower waveform is shown against an expanded, or magnified time scale.

Figure 2.12 Magnifying the event

Putting real figures in this, let us say that from t1 to t2, is 0.5 µsec. Therefore, from Figure 2.13, t2 to t3 is also 0.5 µsec, making a complete sine wave cycle of 1.0 µsec (one millionth of one second).

$$f = \frac{1}{1.0\ \mu\text{sec}} = 1\ 000\ 000\ \text{Hz, or } 1.0\ \text{MHz}$$

In practice, the shape of voltage changes such as shown in Figure 2.13 is rarely quite so predictable, for there are too many other influences to upset so simple a model. It does, however, provide a useful picture of how a practical complex waveform is built up and how it may be dismantled again to analyse.

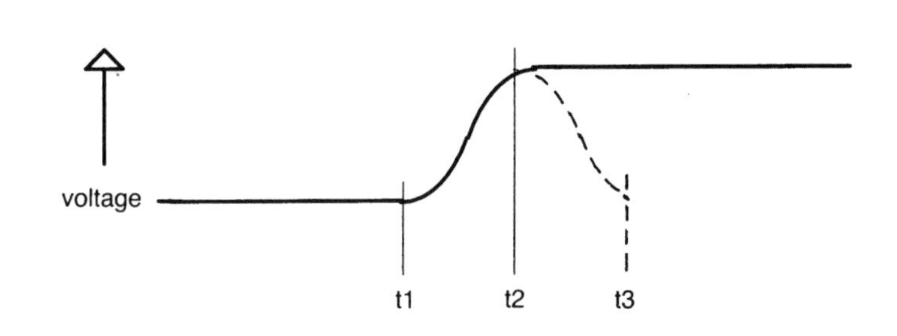

The completed cycle of the sine wave shown in Figure 2.12. Knowing the values of t1, t2 and t3 enables the period of the sine wave to be determined:

$$t3 - t1 = \lambda \text{ secs}$$

which is the **period** of the sine wave. Frequency of the sine wave:

$$f = \frac{1}{\lambda \text{secs}} \text{ Hz}$$

Figure 2.13 The sine form of the voltage rise

CHAPTER 3

WHAT IS VIDEO?

We now look at the most complex signal; video. Video has many features based directly on the points discussed in the earlier chapters. There are also new ones, particularly where colour is introduced but, these will not be dealt with until video as a black and white medium, has been thoroughly investigated. The reason for this is because video was adapted for colour, it was not an original design requirement. To understand colour, we must therefore, fully appreciate the basic black-and-white system.

In our study of audio in telecommunications, we were dealing with a signal of a serial nature; a signal of a continuously varying level; that is amplitude varying with time. A signal has one value of level at any particular point in time. And one that, therefore, converts naturally to an electric signal that may be sent down a two-wire circuit.

A picture signal is quite a different situation.

Sending pictures

The picture in Figure 3.1 has been created by the photographer to achieve a specific response but, as with all pictures, the final outcome depends on the observer. Our eyes never scan a scene consistently: the signals they generate are not the ordered sequential waveform of our hearing. The consequence of this is that picture transmission has quite different criteria, particularly for moving pictures.

To send the picture in Figure 3.1 down a cable will require that it is processed in some organised way. The first consideration is whether this will be in a **parallel** or **serial** form.

The parallel method is a straightforward concept. The picture is broken into individual **pixels**: these are the smallest picture elements that are discernible. Each pixel connects to a signal transmission circuit that, in turn, connects to a picture display where the image will be reproduced as in Figure 3.2

Figure 3.3 shows the analogy where the image is converted into serial form, passed into the cable and transmitted to the display where it is reconstituted. The sequence of image removal, or read-out, is important for what the sensor does; the display must follow exactly. The signal must also carry information about how to do this.

Whether the parallel or serial method of transmission is used depends

Our eyes may scan the picture in a random way depending on such factors as the subject of the picture, how it is composed and the attitude or emotion of the observer.

Figure 3.1 A typical scene

on, above all, how far the signal must travel.

A low resolution image is illustrated in Figure 3.4. It has only 100 pixels and the image is unrecognisable. The quality of the picture determines how many pixels are required. As the parallel method requires a connection per pixel, its conversion to a serial signal form is essential for practical picture transmission.

The transmission circuit now becomes the most significant part of the system for it must transmit the picture *in a given time*. Were it possible to take, for instance, a whole day to send the image then the cable, and circuit, demands are low. Raise the image rate and, over the same circuit, we would have to accept a lower quality reproduced picture. The two are interchangeable, dependent on signal bandwidth. Where we require high quality pictures sent in real time, the circuit requirements rise significantly. Video falls into this category.

The video system

The two video systems most widely used are the European PAL and the American/Japanese NTSC systems. Both are similar in the technology

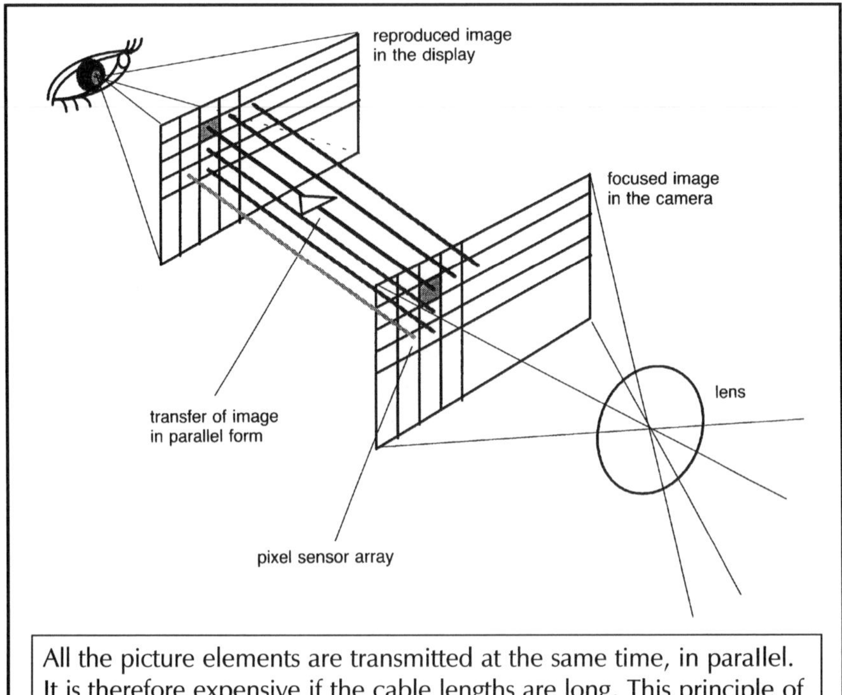

All the picture elements are transmitted at the same time, in parallel. It is therefore expensive if the cable lengths are long. This principle of image capture and transfer is similar to the negative/positive process of film printing where the transfer is done with light. But, unlike film, video uses electrical connections.

Figure 3.2 Parallel transfer of the image

they use, which determines that both have similar picture quality. For reasons of clarity, the PAL system will form the subject of the discussion, but where it becomes relevant, due to the inherent differences, the NTSC system will also be included.

The PAL system uses 625 horizontal lines for scanning pictures. Horizontal line scanning was the method chosen by early television engineers to send moving pictures down a cable. It is in serial form with the image read off left to right and top to bottom.

Figure 3.5 shows the basic scanning system. The video standard is based on 25 frames per second, similar to the 24 frames per second of cinematographic film. About 600 lines of vertical resolution, equivalent to the vertical pixel count, is required for adequate resolution on an average TV screen. The first problem here is the visually annoying flicker rate of 25 frames per second. Film overcomes this by showing each frame twice, doubling the flicker rate to 50 times a second. The eye is much less sensitive to flicker at this rate.

To overcome this problem with video a faster frame rate could have been chosen. But there is a penalty; that of bandwidth.

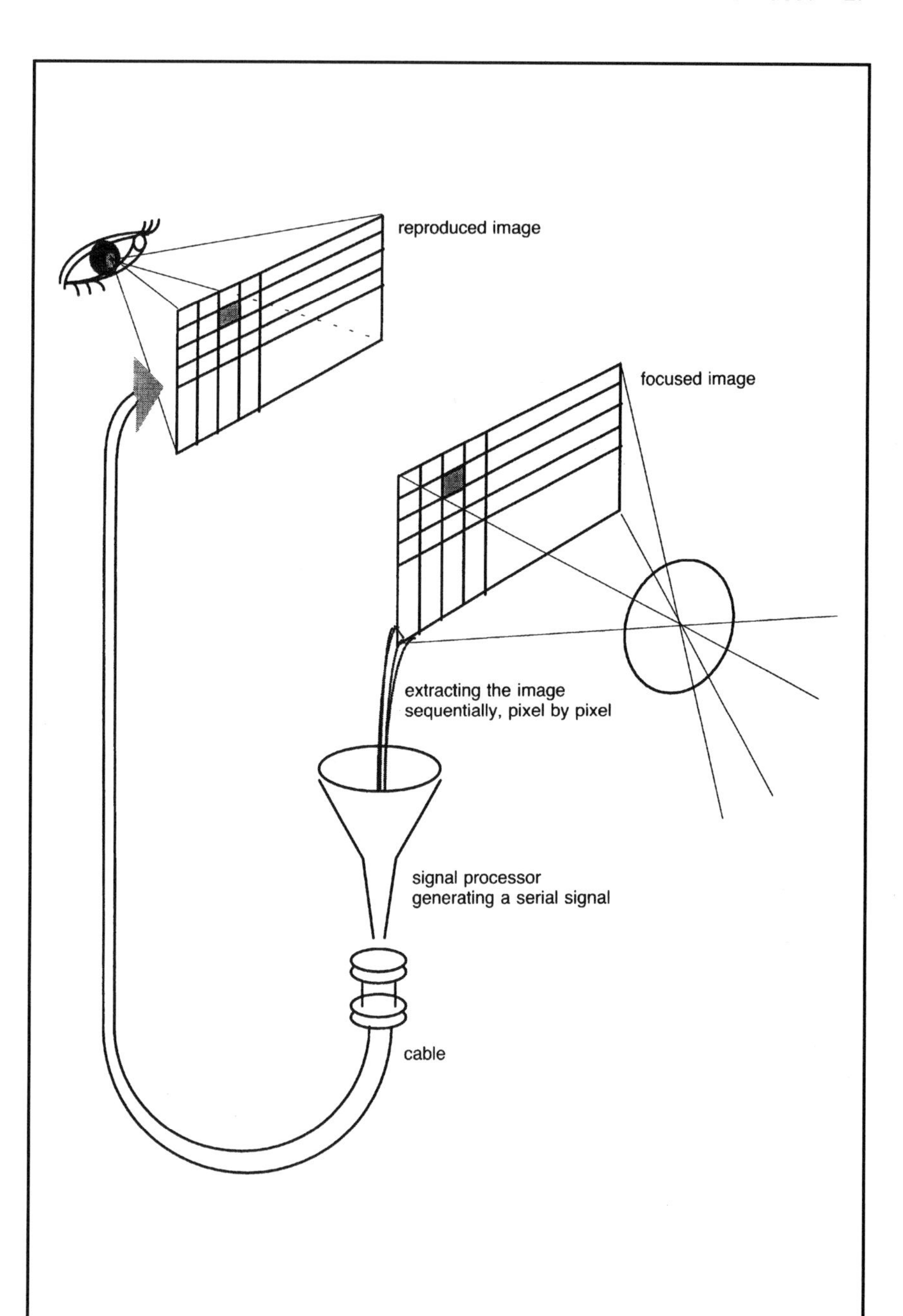

The multiple connections of the parallel system are replaced by a single circuit. The focused image has to be 'dismantled' into a pixel sequence that can be transmitted as a serial signal one pixel at a time.

Figure 3.3 Serial transfer of the image

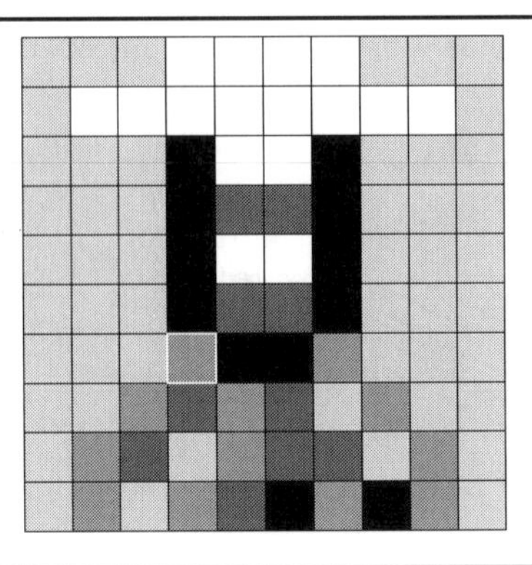

Video uses a CCD (charge coupled device) image capture system. An electrical charge pattern is built up on a silicon pixel array that represents the light pattern of the focused scene. The charge is read off by a processor that converts it to a signal of serial form. A video camera may have up to half a million pixels.

Figure 3.4 10 10 pixel array

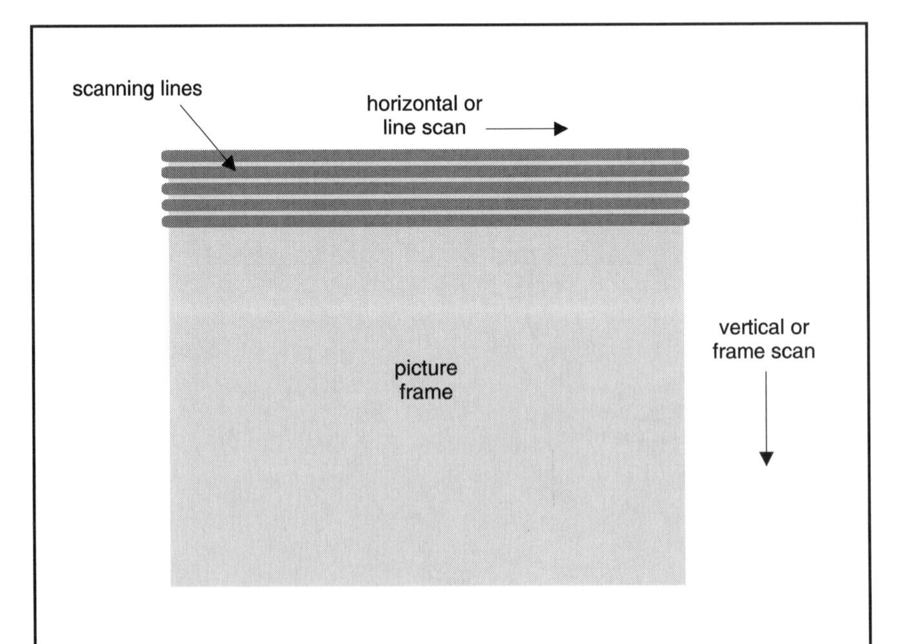

The image is scanned from top left to bottom right. Video uses 25 pictures, or frames, per second.

Figure 3.5 Scanning a picture frame

Let's look at what bandwidth is required for a picture with 625 lines, 25 frames per second and a 4 by 3 aspect ratio.

From Figure 3.6 the number of horizontal pixels can be determined as 833.

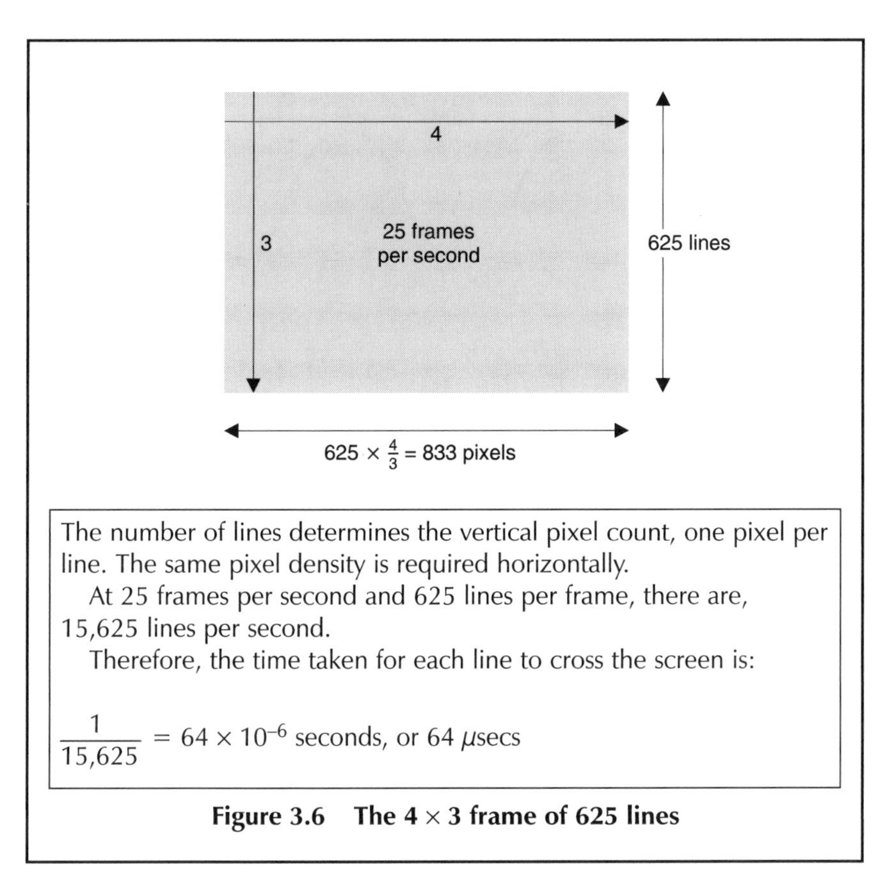

The number of lines determines the vertical pixel count, one pixel per line. The same pixel density is required horizontally.

At 25 frames per second and 625 lines per frame, there are, 15,625 lines per second.

Therefore, the time taken for each line to cross the screen is:

$$\frac{1}{15{,}625} = 64 \times 10^{-6} \text{ seconds, or } 64\ \mu\text{secs}$$

Figure 3.6 The 4 × 3 frame of 625 lines

The finest scene detail that can be reproduced is where alternate pixels are exposed as in Figure 3.7.

Figure 3.7 shows how bandwidth controls the shape of the waveform of the alternately exposed pixels. This is the maximum picture detail achievable by the sensor, and translates into the maximum signal frequency to be handled. The circuit bandwidth is calculated to pass this frequency. Theoretically, the signal will produce a signal step as it passes from one pixel to another but this will be limited to its sine form by the upper bandwidth limit.

The resulting waveform corresponds to 416 cycles per 64 μsec line period. Therefore:

$$\frac{416}{64 \times 10^{-6}} = 6.5 \times 10^{6} \text{ Hz or } 6.5 \text{ MHz}$$

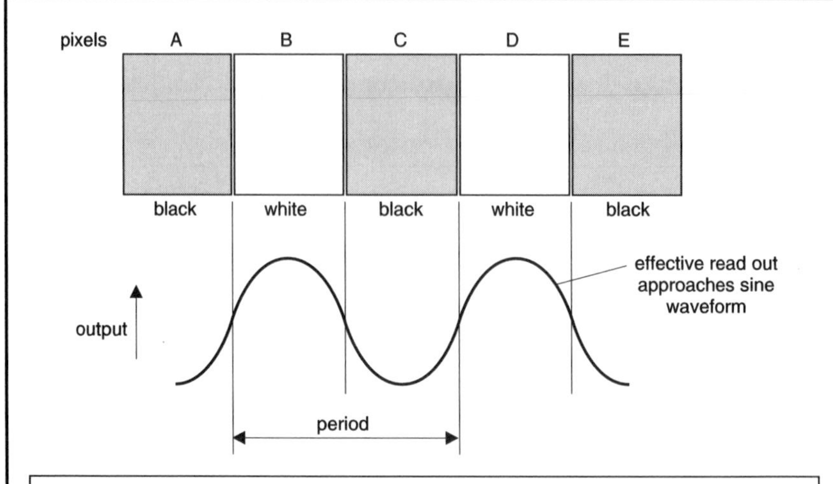

Pixels A, C and E are unexposed, pixels B and D are fully exposed. These represent the maximum contrast of the scene and also its finest detail. When these are read off the signal produced will have the maximum amplitude and resolution available from the sensor. The signal frequency produced from a sensor read-out having 833 horizontal pixels will therefore have a period *twice the pixel size*.

Figure 3.7 Alternate exposed pixels

This is the ideal bandwidth required for 625 lines running at 25 frames per second. As so often happens, other factors come into play that influence the final result.

As we can see, doubling the frame rate to 50 per second to reduce flicker, will require twice the bandwidth. For those interested, calculate the bandwidth for 50 frames per second.

We could halve the number of lines, use a frame rate of 50 per second and retain the same bandwidth. The flicker will be overcome but, of course, resolution will be halved.

Why the concern over bandwidth? Well, bandwidth is space, and space in the transmission spectrum is cost. Increasing bandwidth will also increase complexity and impose design restraint thereby causing additional cost penalties. It may even push the system beyond the technology available to build it.

But we have still to resolve the flickering of twenty-five pictures per second if we are to watch in comfort.

A neat solution, using the manipulative ability of electronics, is to break each frame into two fields. By retaining the original frame rate of 25 per second for 625 lines we retain the same bandwidth and resolution. Each frame is now made up of two fields, each with 312½ lines.

The whole structure of the video picture is shown in Figure 3.8. Each line is traced out with a continuing movement down the picture. After every line there is a rapid **flyback** to commence the next. The fields are numbered 1 and 2 to identify them, field 1 starts with a half line, field 2 ends with a half line. On completing a field, the scan returns to the top of the picture to start the next but, this time, displaced by half a line.

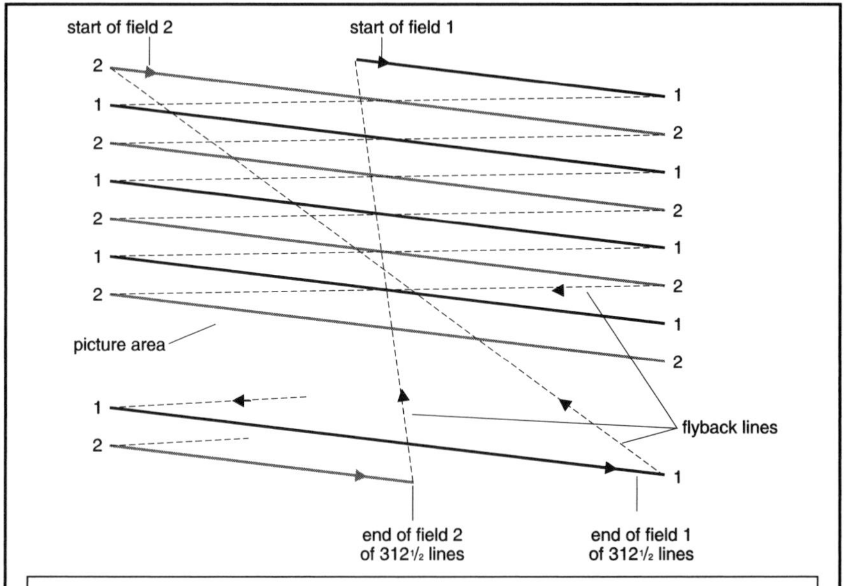

The picture display follows the same pattern as the image sensor with a scanned area larger than the actual picture. Each line commences outside the viewed area, a number of lines at the top of each frame are likewise unseen. Therefore, the number of 'active' lines is less than the total used.

Figure 3.8 The interlaced frame

The process repeats every two fields, making one frame. The frame rate becomes the system lowest, at a base frequency of 25 Hz. Note how the line rate frequency of 15, 625 Hz is a multiple of 25 (that is, 25 × 625), making the resultant a complex waveform based on these two frequencies. Where scene detail gives rise to a signal with elements approaching the maximum of 6.5 MHz., as in Figure 3.7, then the waveform becomes very complex indeed.

So we have seen how scanning the picture in this way, produces a serial signal that can be sent down a two-wire circuit. At any instant in time, the signal has one value only, that of the brightness of a single pixel. We can say that the picture has been coded into a video signal.

When the bottom of the picture is reached, the scan returns to the top during **field blanking**. Field blanking is ten lines in length but although blanked, these lines continue the scan process right through field flyback (Figure 3.8 does not show this for reasons of clarity). The length of field blanking is therefore:

64 μs × 10 lines = 640 μs

One can therefore consider blanking as a picture framing or black mount. Many displays cut off, and do not show the picture edges, for the active scan size actually exceeds the physical size of the screen. Where the full scan is seen, however, blanking ensures a very clean edge to the picture.

Another implication of this is that, because 10 lines of each field are blanked there are only 600 or so active picture lines in a frame. This produces a saving in information and therefore bandwidth.

The timebase

The picture display must accurately follow the original image scanning, whether this be from camera or any other video signal generator. This is **synchronism**. Both must start at exactly the same time, maintain step, and finish at the same time. Should they get out of step the display will not know which part of the signal is which, will become confused and fail to reconstruct the picture.

To ensure image and display are kept synchronous a code is added to the picture waveform to keep the display system updated.

The whole video structure is based on time. Time is a relatively easy parameter to measure electronically; our quartz watches are excellent examples of this. Consider a camera pointed at a scene, its electronic clock or **timebase** instructs the processor of the CCD sensor chip on which the image is focused, to start the sequence at line 1. However, before the camera actually commences scanning the picture, a marker is placed in the waveform which instructs the display to start its scan and await the arrival of the picture.

When the first line is complete another marker leaves the camera, telling the display to flyback and start the next line. Then, at the bottom of the picture at the end of the field 1, a sequence of marks is generated by the camera saying the last line of that field has been sent and to return to the top of the picture and await the first line of field 2.

We can see from this that there are two principal timing requirements, the frame rate and the line rate. These are related, as we saw above, and it could be argued that it is not necessary to send timing information at each line. After all, modern clocks are very accurate and once started, the display could time itself with its own clock. Synchronism, as has been said, is fundamental to television and to understand why slaving the display so accurately is essential. Let's do a little calculation.

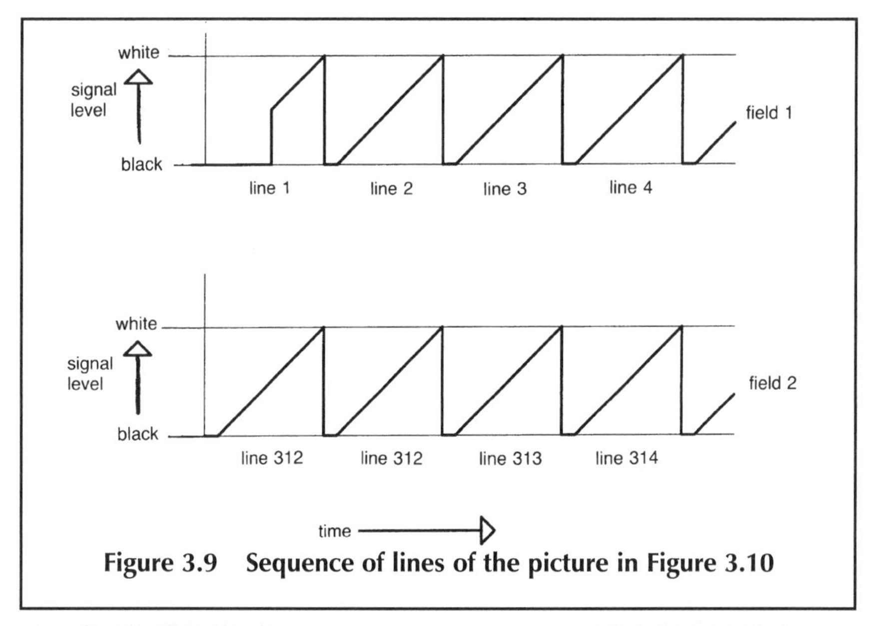

Figure 3.9 Sequence of lines of the picture in Figure 3.10

Figure 3.10 A ramp test signal

The picture is of a **ramp** test signal; it appears as shading from black to white, left to right. The name originates from the shape of its waveform. The waveform is of the top eight lines of the picture. Line 1 is a half line, starting the scan of field 1, line 312 starts field 2. The waveform shows the signal level rising from black, reaching white, then rapidly returning to black. A gap exists before the next line, called the **line blanking** period where all picture information is replaced with black.

In a one hour programme, there will be:

625 × 25 frames × 60 seconds × 60 minutes = 56 250 000 lines scanned.

If the receiver clock has an accuracy of 1 part in 100 000, typical of a quartz clock, then this can gain or lose 562 lines during the one hour programme. Without regular timing markers this error would amount to almost a whole frame out of sync by the time the programme draws to a close. Even with a marker every frame, such time keeping is still not good enough to ensure the verticals in our picture do not become skewed by a line timing error.

Timing is therefore very important indeed and the scanning, or time-base, clock in a modern television receiver still needs constant timing updates to maintain the perfect synchronism required for good quality pictures.

What are these timing markers? In Figure 3.9 we see the waveform of the first four lines of fields 1 and 2. Between each line is a period of black, and it is here that the timing markers are placed. These are the **line sync pulses** and Figure 3.11 shows what they look like.

Figure 3.11 shows how picture and timing information are added together to complete the video waveform.

The illustrations show an idealised representation of the waveforms. For instant, there are drawn instantaneous changes of level, our discussion on bandwidth pointed out the difficulty here. There is an instant return from white to black at the end of the ramp. Likewise, the sync pulses are shown with vertical edges. Such is not the case in practice but in this diagrammatic style it is quite usual to see signals represented in this way.

The scanning structure is, of course, invisible to the viewer, only the picture elements that are brighter than black will be seen. Therefore, to ensure that the correct values of black and white level are reproduced and no artefact of the process is seen, the standards for signal levels and timing must be adhered to.

The video waveform

The ramp waveform in Figure 3.11 is a signal of maximum amplitude in that it includes the black and white level, plus sync level. These are the maximum limits available to the picture maker and any excursion beyond these is prohibited and will normally be removed.

The complete signal adds up to a maximum of one volt. But not all pictures possess a full range of tonal values and not every waveform will make full use of the capacity available to it. For example, not all pictures have black in them, nor do all pictures possess a white. The interesting case is where there is no picture at all – a black screen.

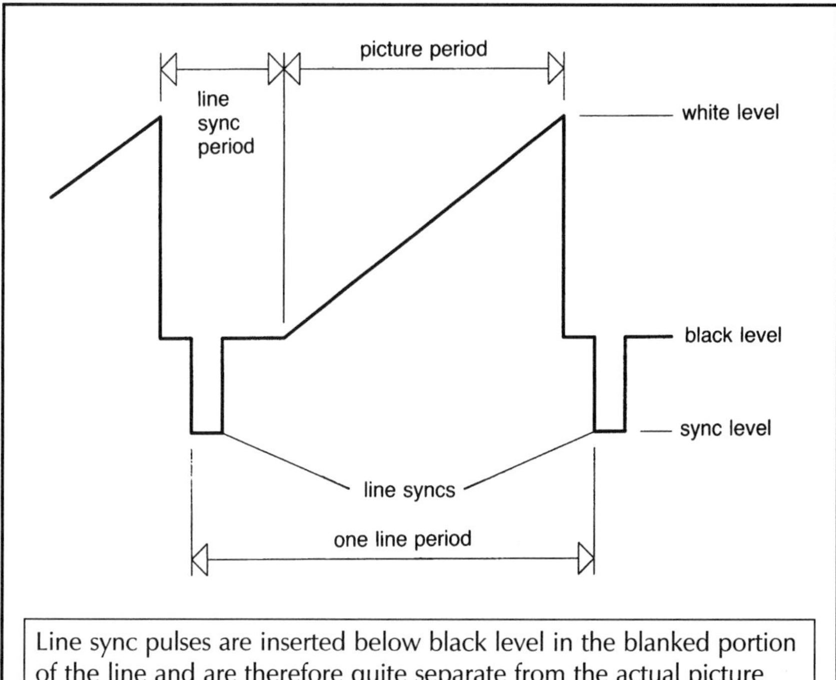

Line sync pulses are inserted below black level in the blanked portion of the line and are therefore quite separate from the actual picture signal. They take the form of unseen voltage pulses outside the picture area and below black. Electronics and video make unreal concepts, such as blacker than black, perfectly possible when actual voltage values are added to the waveform.

Figure 3.11 Ramp waveform and line sync pulses

Black is a common situation; there are often moments when a picture fades to black. However, this does not imply a lack of signal, far from it: all the timing requirements will still be present. The display must always be ready to show the fade-up from black to the next picture. Sync information *must continue unbroken* at all times.

Figure 3.12 shows the standard video signal with the voltage values.

Black and white are the picture limits of the video system. The system will not allow any excursion below black. Although this may appear an unreal concept, 'blacker than black' is quite real in electronic terms for any value of signal may be represented by a specific voltage. Fixing black at 0 or 0.3 V does not inhibit our video system. Indeed, as we have already seen, outside the picture area the timing information is carried below black. And so we place a limit to picture information going below black for if we do not, we run the risk of interfering with the timing process causing the display to interpret this as sync pulses and become confused.

At black level, therefore the picture is clipped in the camera. A similar limit is applied at white level, often known as 'peak white'.

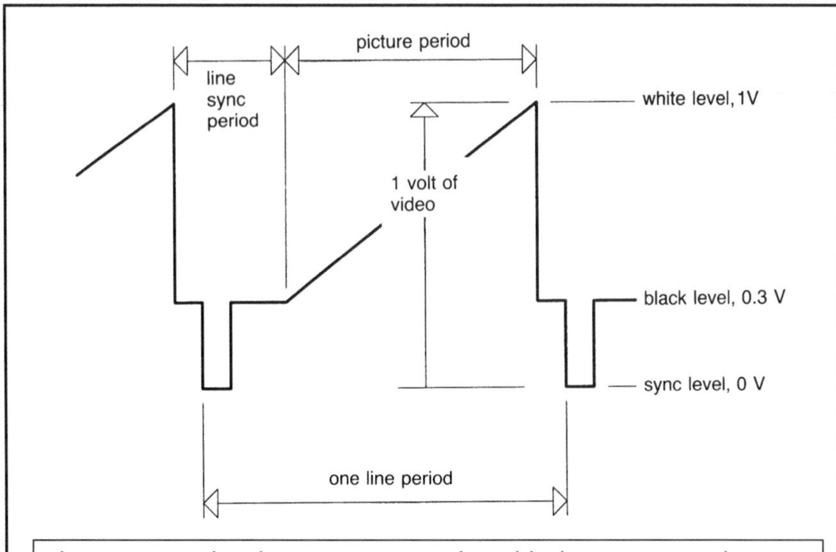

The ramp signal is shown running up from black at 0.3 V to white at 1.0 V. Sync level is the reference at zero volts. Black may in some cases appear at zero volts, making sync level –0.3 V and white 0.7 V. The picture however, always occupies 70% of the waveform and the sync 30%. The term 'one volt of video' refers to the whole signal.

Figure 3.12 One volt of video

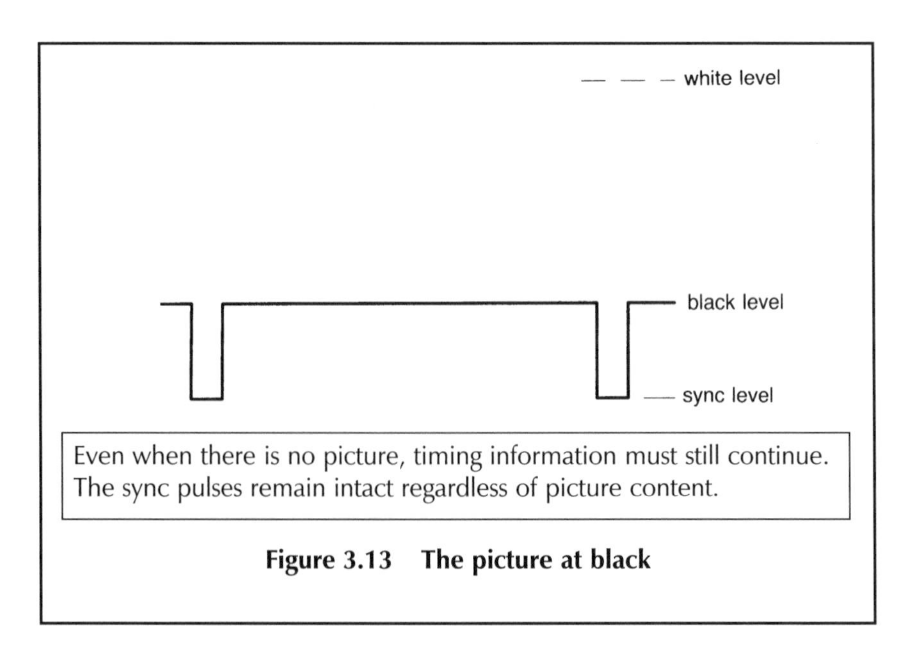

Even when there is no picture, timing information must still continue. The sync pulses remain intact regardless of picture content.

Figure 3.13 The picture at black

CHAPTER 4

MEASURING VIDEO SIGNALS

Why do we measure? Signals must fit into the system that carries them, if not, then the situation shown in Figure 4.1 will arise. All transmission systems, whatever the medium or method used, have maximum signal limits. Where signals exceed this maximum, **distortion** of the signal and the information it carries, will occur. The form of distortion will depend on the system and how it has been misused. How much actual degradation of information takes place depends on how resilient the transmission system is.

The left hand figure passes through the doorway easily. The one on the right is too tall. If a signal is too large the system will distort it and it is also likely that damage to other signals will occur where the system handles more than one at a time.

Figure 4.1 Insufficient headroom

Signal level exceeding the standard limits is the most common distortion. Time is also subject to distortion. Signals take specific times to pass through systems and where these **time delays**, as they are called, are altered, we get signals arriving at the different times. Remember, that video is a time-based system; we will see the implications of this later on.

Measuring the signal

Video is generated and sent as a voltage; we therefore require a **voltmeter** to measure it. Voltmeters are usually associated with steady-state conditions but video is not steady-state, it is a constantly varying signal of a moving picture. It also has the two distinct elements, picture and synchronising pulses, each needing their own specific measurement requirements. We must therefore have an instrument that allows us to 'see' the signal voltage as a waveform and to offer the ability to observe and

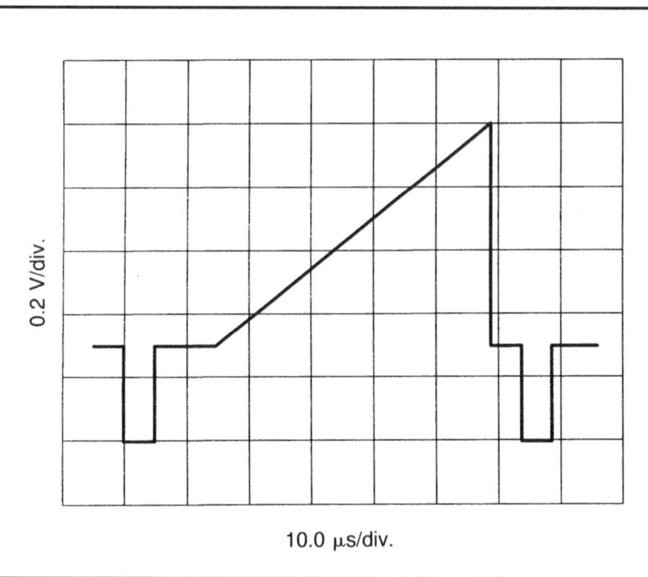

The graticule is a measurement scale of both level and time. Measurement is done by shifting either the graticule or waveform until the section of interest is aligned against a graticule line. The scales here are: vertical, one-tenth of a volt per division and horizontal, ten microseconds per division. Note that the calibration is not absolute; we can choose where the scale is placed. As shown, the bottom of line sync and highest picture level are 5 divisions apart, which corresponds to 1.0 volt. Horizontally, the time scale shows line sync as 5 μs. wide and the distance between syncs as 64 μs. Note how the 1 cm squares are not sufficient for accurate measurement. Additional increments of 0.1 cm are usually added.

Figure 4.2 Placing a graticule over the waveform

measure selectively.

We measure by placing a scale against that to be measured. A ruler measures in one dimension; video has two dimensions: level, or amplitude, and time. The scale is called a grid, or **graticule**, as in Figure 4.2. Here, the two dimensions appear as X, horizontally and Y vertically.

Figure 4.2 uses a typical measurement scale allowing measurement of all parts of the waveform to be done quickly. It is possible to alter the scale of measurement, not by changing rulers, as it where, but by changing the amount we display. Figure 4.3 illustrates how two lines may be studied by doubling the observation time.

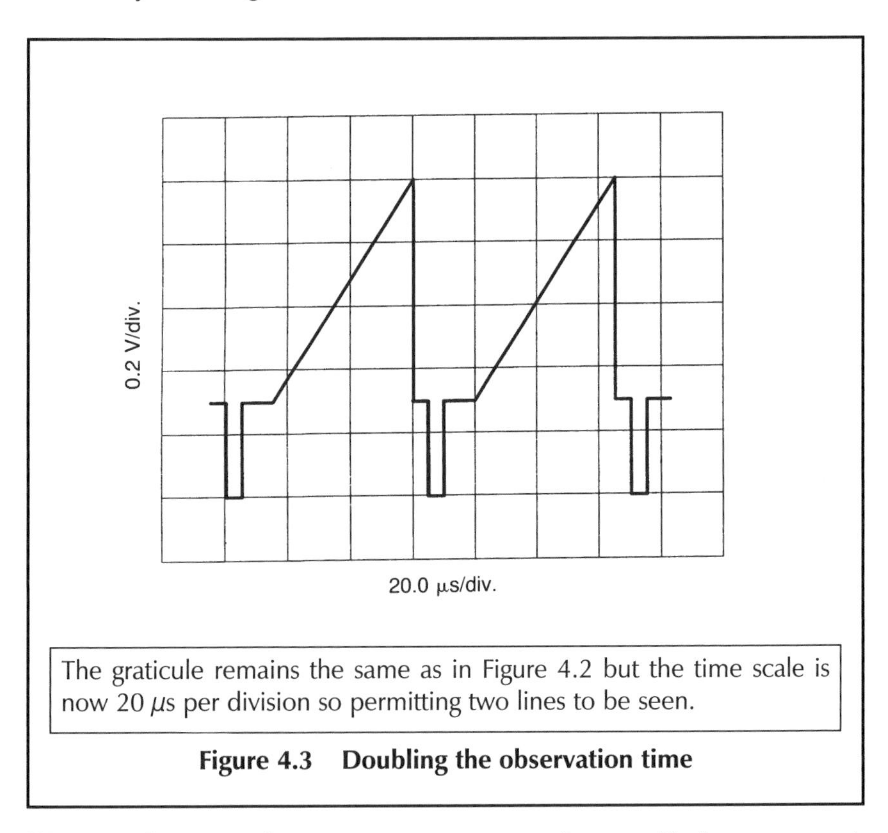

The graticule remains the same as in Figure 4.2 but the time scale is now 20 μs per division so permitting two lines to be seen.

Figure 4.3 Doubling the observation time

We must also state where our measurement points are. To do so we must establish the standard terms for various parts of the waveform, as in Figure 4.4.

The waveform monitor

The means to see a waveform has, traditionally made use of the **cathode ray tube**. The principle is very similar to that used for picture displays but over the years there has been a divergence between the two designs. The waveform monitor WFM has now become the established method of

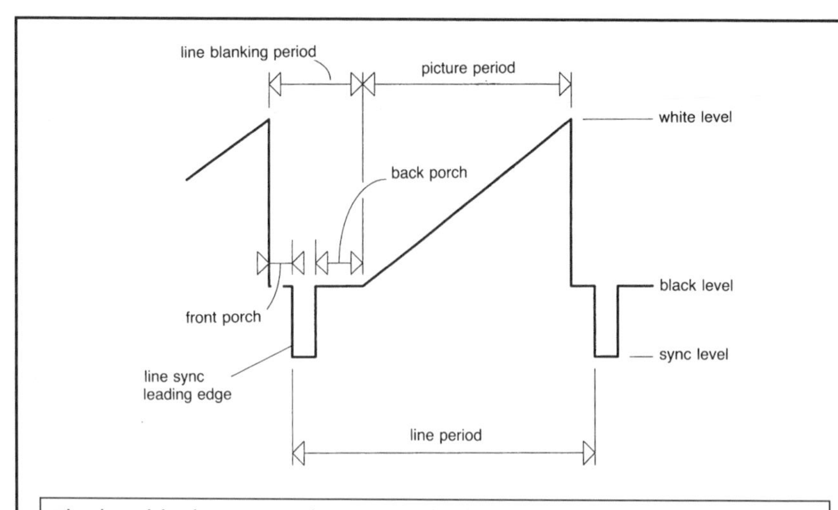

The line blanking period contains the line sync pulse, with **back and front porches**. The porches also represent video black level that separates the sync period from the picture. The line period may be measured between any two similar points, but it is usual to measure from leading edge to leading edge, of line sync.

Figure 4.4 Terms in common usage

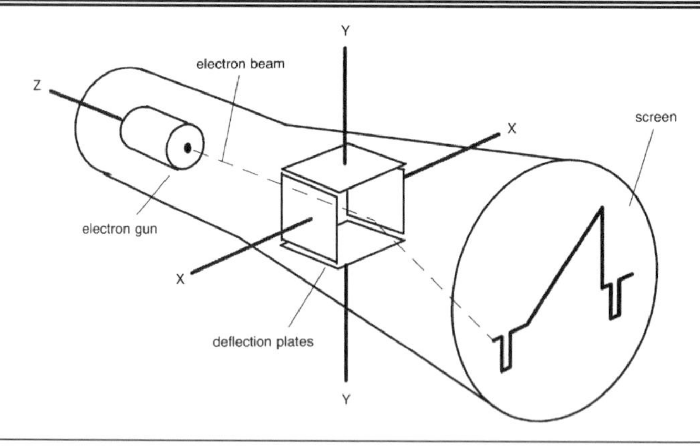

The electron beam, travelling in a vacuum, strikes the phosphor coated screen causing light to be emitted. A voltage applied to the deflection plates results in electrostatic forces that deflect the beam. Plates YY carry the signal waveform, controlling the beam vertically, whilst XX carry the horizontal timing control. The Z connection to the electron gun allows the beam to be turned off during its return flyback.

Figure 4.5 The cathode ray tube (CRT)

displaying waveforms and Figure 4.4 shows the basic principle of how the WFM uses the CRT.

Figure 4.5 shows a basic CRT as used in waveform monitoring. Electrons, travelling at high velocity, are formed into a beam that, on passing under the influence of the voltage field between the deflection plates, becomes deflected. The signal must be raised to hundreds of volts for the beam to be fully deflected. The simplified circuit in Figure 4.6 shows how the signal drives the Y deflection plates in antiphase, that is, one plate swinging positive whilst the other swings negative.

Other forms of display are also in use, for instance, LCD (liquid crystal) techniques. Standard picture monitors may also double as WFM's by proprietary video adapters. All these variations use common parlance to describe the various functions. For the purpose of our discussion, the CRT-based method will be used.

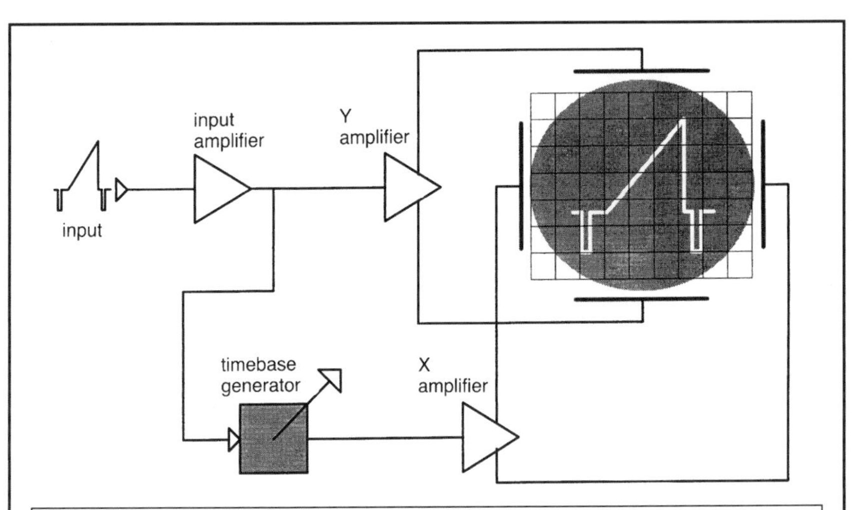

The input to the WFM is the video signal to be measured. The input amplifier passes the signal to the Y amplifier driving the Y plates controlling vertical beam deflection.

The input amplifier also presents the signal to the timebase generator where the line syncs initiate a signal for the X plates that moves the beam across the screen at a steady rate.

A timebase generator, running under the control of its internal clock synchronised to line sync pulses, produces the timing waveform for the X deflection. This action continually repeats with the horizontal deflection and flyback synchronised to the video signal.

The timebase clock shown has a variable time rate (as indicated by the arrow), providing different time lengths over which the signal may be observed. In Figure 4.2 the display is shown at line rate whilst Figure 4.3 is at twice line rate as determined by the clock rate.

Figure 4.6 Principle of the waveform monitor

There are many variations in the design of WFM's. The graticule is designed for video with level and time scales. There may also be, as seen here, a read-out of the main functions.

Figure 4.7 Waveform monitor with ramp test signal

As we delve deeper into the complexity of video, so the rather simplified view of the waveform monitor seen so far develops and adapts to the requirements of a practical system.

Black level

Chapter 3 made reference to 'black', that value in the picture where no light leaves the screen. We must now define this more carefully for it is crucial that the picture display knows exactly what this value really is.

We have seen how the video signal is made up of two distinct parts; the picture and the technical. The technical part is the sync information lying below black level. Black level not only provides separation between the two, but offers a picture black reference.

In Figure 4.2 the waveform shown was stated as having no absolute value, but points out that it is quite possible to measure the amplitude overall and its constituent parts. The maximum level, or peak white is measured as 0.7 V above black. The absolute value of black is not stated and it is this that the picture display requires.

Any picture, to be reproduced accurately, must have its tonal values defined. It is insufficient to state these as relative values, i.e. that a face is brighter than the wall behind it. The level of brightness of the face must be converted into a signal value and, on arrival at the picture display, has then to be reproduced at the correct brightness. To do this we refer everything to black.

Now black may not exist in all pictures but it can be defined as a value in the video signal and a voltage value given to it. The value of black level is hidden in the sync period of the signal where the display can use it as a reference and from this construct all the tonal values in the picture.

As the signal passes through various circuits and systems, it may become subject to extraneous influences that result in the value of black level shifting. The value of black is very precise. Quite small variations up or down can have very considerable impact on the picture reproduction. For instance, an inadvertent shift of picture black during signal transmission or processing by only a few percent is far more critical than a change of overall level. The value of picture black must therefore be retained accurately.

The video standard states that reference black will always be that of the back porch, the portion of picture blanking that precedes the picture. Having established this, it becomes a simple procedure for the picture display to locate this point in the waveform. It is then able to base the actual picture on this, placing all scenic tones at their correct values. Note that our picture may not contain black as defined in the waveform, but this does not alter the principle of sending a true video black level as a picture tonal value reference.

In practice, black level can be allocated any convenient voltage. As long as peak white is at +0.7 V and sync level at –0.3 V, *with respect to the reference black level*, the signal will still comply with the standard. Practical circuits, whether transmission or recording, may cause the absolute value of black level to change. Over practical distances of signal transmission, even a hundred meters of cable, interference from other apparatus may affect our signal, the most common form being that induced by power circuits. Picture displays, and other processing equipment must have, therefore, the means to recover the reference black.

The process is known as *DC restoration* and it is carried out at every line in the back porch. The term 'DC restore' means to restore the DC (direct current) component of the video signal. This action has become standard practice in signal processing and transmission.

What the waveform monitor shows

We have now put in place all the background to make measurement possible. There is a voltmeter in the form of a waveform monitor; the WFM, measuring signal level in voltage. We have ascertained the parameters of a standard signal and its references. Black level is established as the reference from which to measure picture levels, although overall amplitude

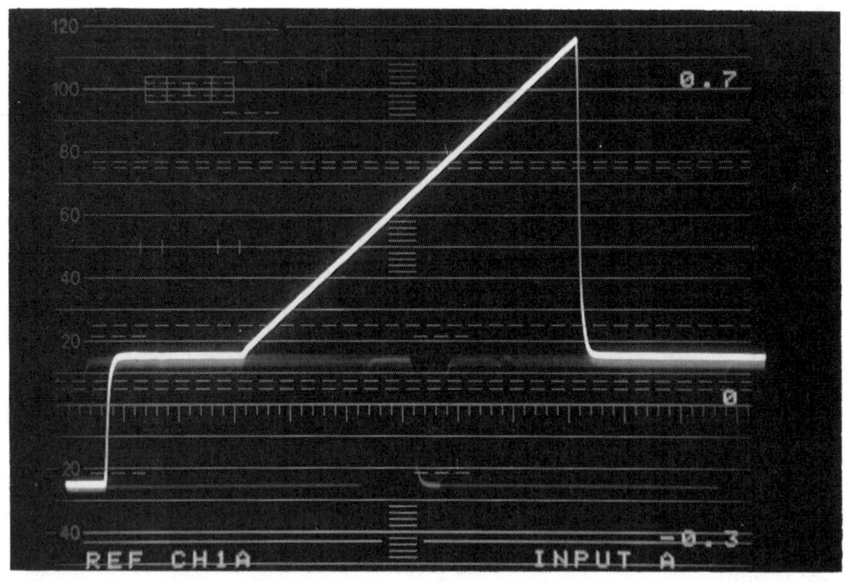

(a)

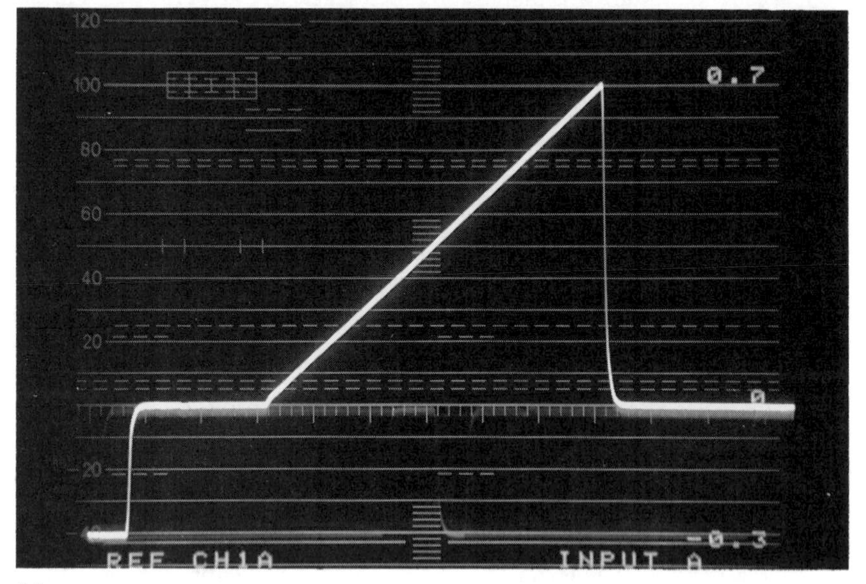

(b)

These pictures show the relative shift of waveform against graticule. (a) is positioned too high; (b) has black level positioned correctly against the graticule. Peak white is at 0.7 V and sync at (–0.3 V, as indicated by the read-out. The graticule shows levels in 5% steps with 1% increments at the centre. A single time-scale is shown at black level and relates to the X or horizontal deflection control setting in (μ/division. Other markings are aids towards specific equipment set-up.

Figure 4.8 Measuring the video signal

is measured from the tip, or bottom, of sync pulses to the maximum value of the picture.

It is usual to state 'one volt of video' as the amplitude of a standard test signal, in the same way that actual pictures do not always possess black, they do not always possess peak white. One volt measures from sync level to peak white, whilst measuring from black level is generally carried out for picture-only information. Thus there is a distinction between measuring for test or engineering purposes and operational requirements.

Whichever is used, we need the means to position, or shift the waveform against the graticule. Figure 4.8 shows what this means and Figure 4.9 shows how its done.

The complete WFM in Figure 4.10 is the basic layout and all the principal controls are shown. The timebase is switchable between different time values, the main ones are line and field, often abbreviated to H and V. For instance, 1H is one line, 2H is a display of two consecutive lines, but this does not quite explain what one is actually seeing. With the 1H mode, the timebase triggers the X deflection every line, resulting in a display of all lines superimposed. Figure 4.11 shows how this appears and for comparison, a display of just one line. Note how the latter is less bright for it only has one line's worth of energy every 1/50th of a second, whereas the former has a whole field of 312½ lines over the same period.

Exactly the same applies when considering the field display. Selecting V shows one field, 2V is of both the fields making up one frame. A 2V display is as shown in Figure 4.12.

Magnification of the display is very similar to using an optical magnifying glass. The shift controls move the expanded waveform until the desired part is placed in the graticule. Magnification rates of 5 times to 25 times are common. As only a portion of the waveform now occupies the screen the image is theoretically less bright, a situation rectified by increasing the display brightness.

Brightness of the display is an operation control, made necessary by the different ambient lighting conditions encountered in practice. To compensate for brightness variation brought about by changing the timescale and magnification, an automatic system is usually incorporated into the instrument. It is only possible to increase brightness to that allowed by the screen. Attempting to go beyond this point is liable to distort and make the display less sharp.

Timing and system sync

Video is a time-based system. The information about picture tonal values is sent as a voltage broken down into a sequence of lines and fields. Information about which part of the picture is which is therefore part of a timed sequence. The video standard sets out the timing values of the signal so that any standard picture display will function with picture generating apparatus designed to the same standard.

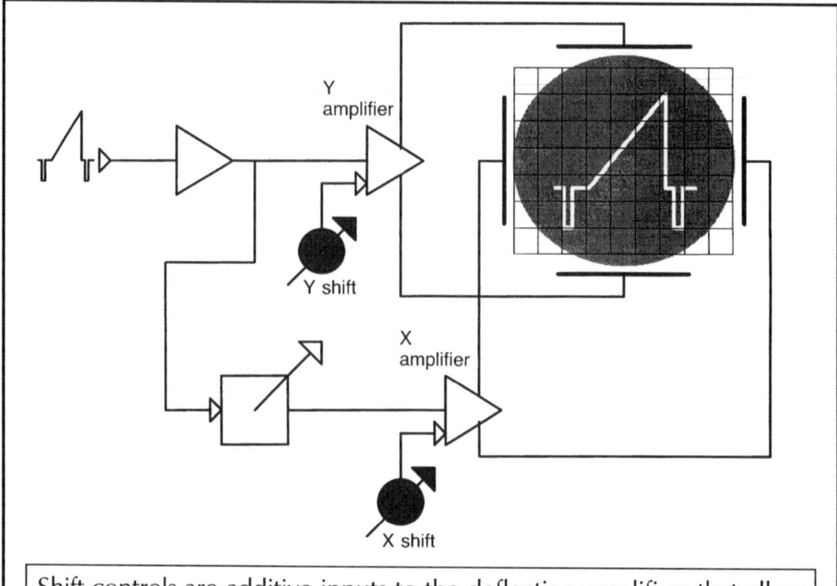

Shift controls are additive inputs to the deflection amplifiers that allow variable DC (positive or negative) to be added to both the signal and timebase deflection drives. They make the display position adjustable against the graticule.

Figure 4.9 Positioning the signal against the graticule

Timing standards must be as rigorously adhered to as amplitude values. For example, two cameras linked to a studio video mixer or switcher, must produce signals in synchronism. If not then, as the switcher cuts or mixes the two pictures, confusing synchronising information will be sent out.

We have already seen the importance attached to the signal's sync pulses. They carry the timing information, telling picture displays, in studio and the home, when pictures start and end. Should our two cameras produce pictures that are not synchronous, then conflict will arise as to which camera the display is to follow.

We have, therefore, two criterion; timing accuracy of the signal, the period (or time length) of lines and fields, and timing synchronism between picture sources. The WFM deals with both of these but in different ways. Line and field period is measured against the time period generated by the WFM's timebase. When time synchronism is to be measured we must choose one of the cameras to be master and compare the timing position of the other against it.

It is the latter situation that provides the WFM with a large part of its duties. Signal time values are a standard and as such are designed into the picture generating equipment, whatever that may be. Such parameters are unlikely to vary unless a fault condition arises.

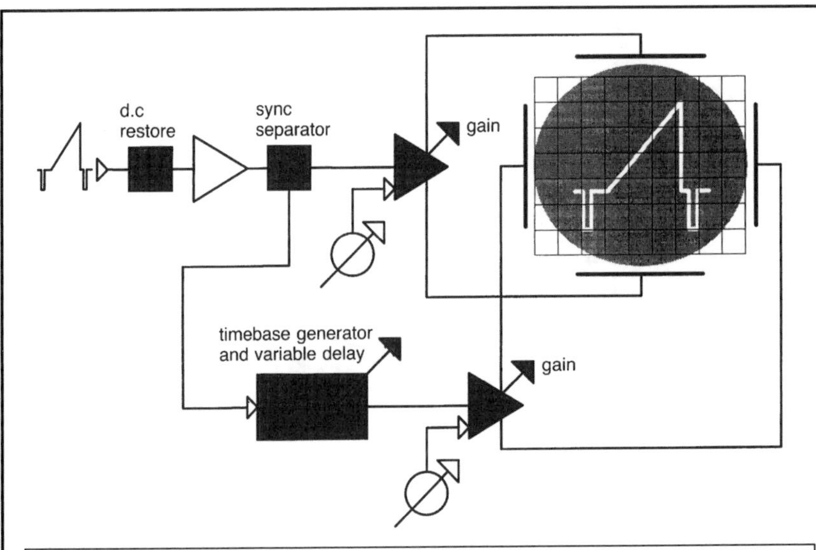

The dc restore circuit at the signal input gives black level a precise value and places black at the same point on the graticule regardless of how the picture values change. The sync separator is a sophisticated circuit that provides precise timing from the sync pulses for the timebase generator. The timebase generator offers variable delay of the Y deflection so that any particular line may be displayed. The deflection amplifiers have variable gain, (variable control of amplitude) to provide the detailed study of signal level.

Figure 4.10 The complete waveform monitor

On the other hand, synchronism is a system feature, with operational requirements. Systems are designed for specific work, e.g. the design of our two-camera studio must ensure that both cameras are synchronous at the switching point. Remember, signals travel at a finite rate, although this approaches the speed of light, it is significant when compared to line rate. Cable lengths now become important. Should one camera be at the other side of the studio on a long length of cable, its signal will be later than one placed closer on a shorter length.

The problem is resolved with the WFM by comparing the cameras to its timebase, or by comparing one camera to the other, as in Figure 4.13.

However, Figure 4.13 does not tell the whole story. If the WFM is selected to each switcher source in turn, each signal will be placed on exactly the same part of the display; the WFM DC restore, sync separator and timebase would see to that. Although signal level differences would be seen, timing differences would not. Figure 4.14 illustrates how this is overcome.

So how close must timing be? Putting a figure on this depends on how critical the actual system is but at this basic stage of the discussion, we

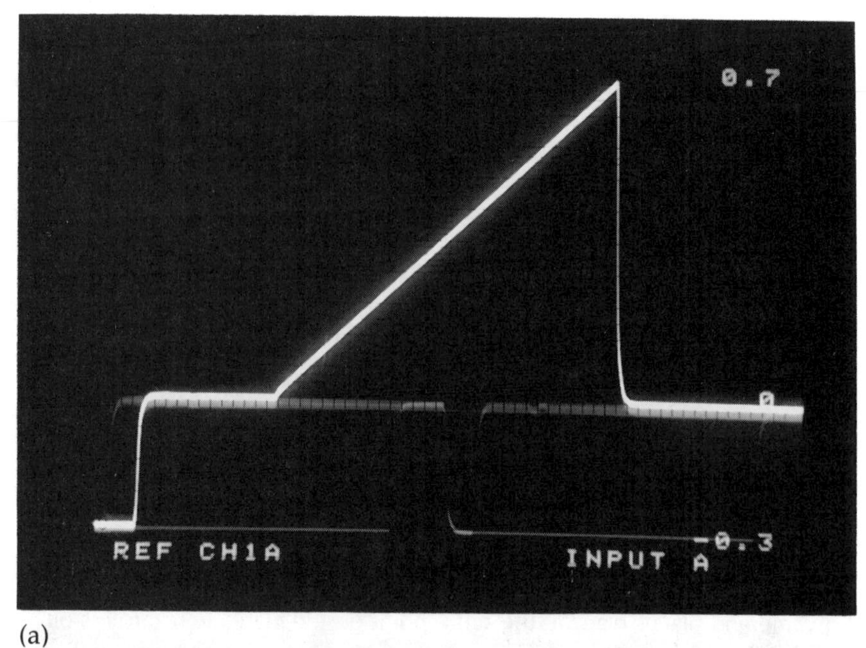

(a)

(b)

All lines are shown overlaid in (a); this is the 1H display. Note there is a spurious signal at black and sync levels. These are the black lines and field sync pulses of the field sync period. In (b), only one active picture line is shown. In both pictures, the graticule light is turned off to show the effect more clearly.

Figure 4.11 Line rate and single line displays

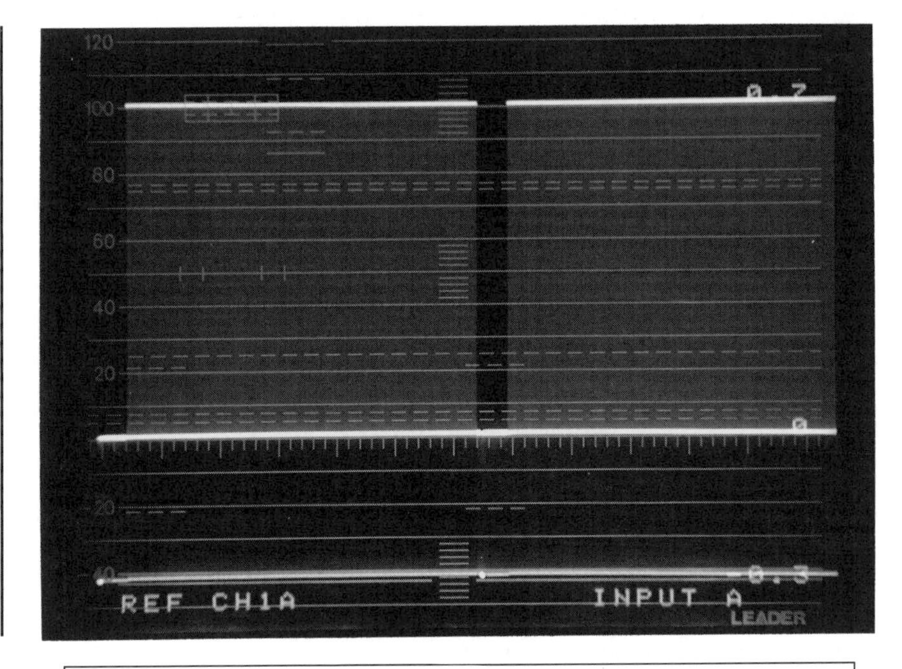

The two-field (2V) waveform consists of all lines shown sequentially but are so close together that they lose individual identity. The black lines of the field sync period are clearly shown. The line sync appears as a continuous line but it is possible to see the brightening of the field sync pulse. The time-scale is 1/25 s, or 20 ms.

Figure 4.12 Field rate display

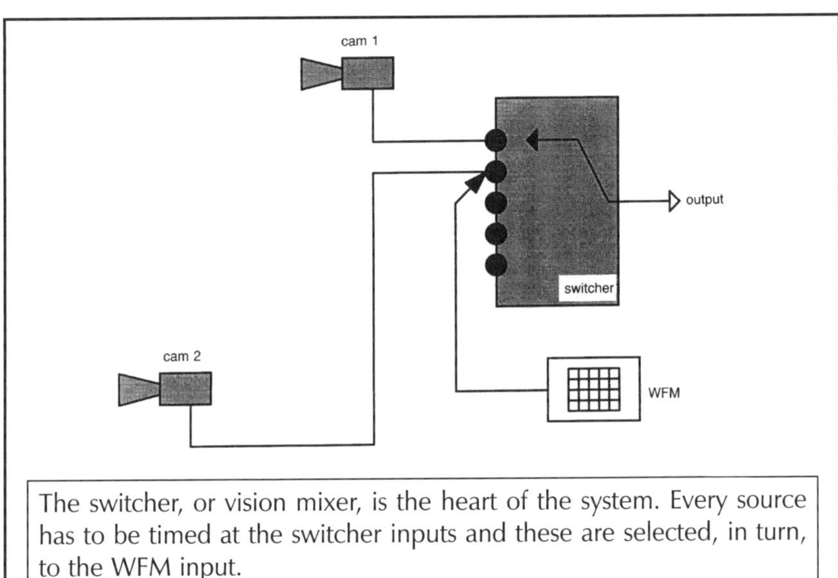

The switcher, or vision mixer, is the heart of the system. Every source has to be timed at the switcher inputs and these are selected, in turn, to the WFM input.

Figure 4.13 A simplified studio system

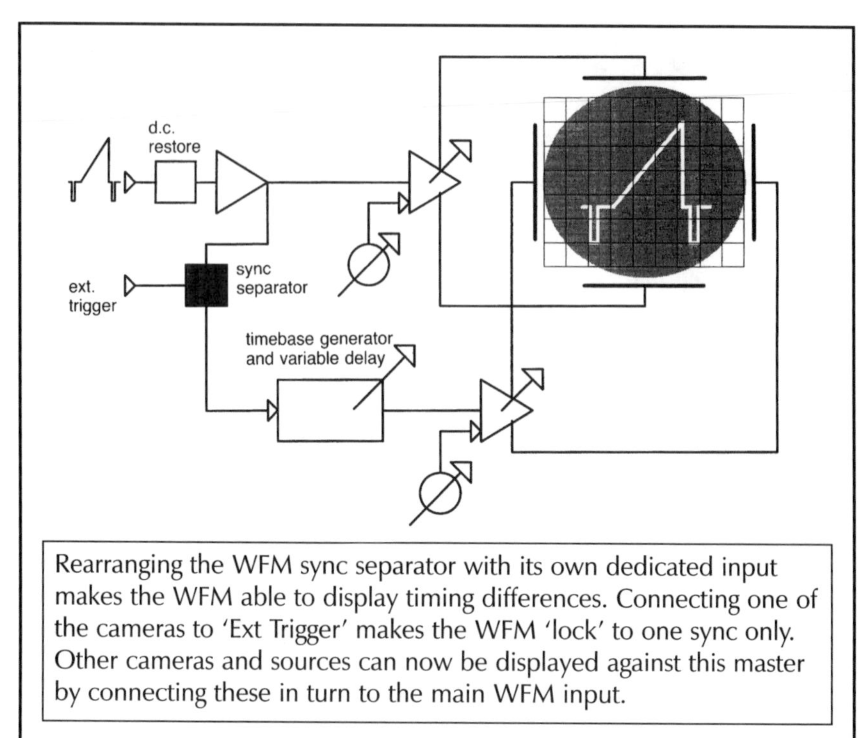

Rearranging the WFM sync separator with its own dedicated input makes the WFM able to display timing differences. Connecting one of the cameras to 'Ext Trigger' makes the WFM 'lock' to one sync only. Other cameras and sources can now be displayed against this master by connecting these in turn to the main WFM input.

Figure 4.14 Synchronising the WFM

can say that timing accuracy should be better than 0.5 µsec. Figure 4.15 shows typical timing measurement.

This rather simplified view of timing leaves much still untold but in subsequent chapters the subject will be developed. And, as colour is drawn into the discussion, we will see a whole new dimension of the video system.

'Scopes v. WFM's

At this point it is worthwhile looking at how the WFM differs from the conventional waveform display of the **oscilloscope**. 'Oscilloscope' often gets shortened to 'scope' or CRO (cathode ray oscilloscope). In principle there is little difference between the two: both show voltage vertically and time horizontally. Indeed, the 'scope is often used in place of a WFM. It is in the operational convenience that the two are distinguished.

With a more specialist instrument, dedicated controls and circuitry become standard. Because the WFM is specifically designed for video, it uses video sync separation to achieve reliable triggering at line and field rates. There is the ability to delay the display for a specified time to study any part of the scanning sequence. For example, the field trigger starts a

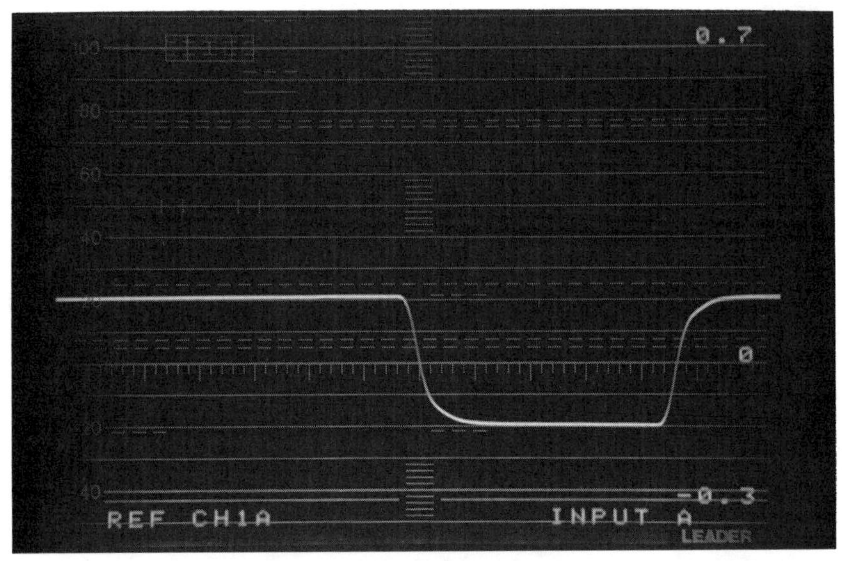

(a)

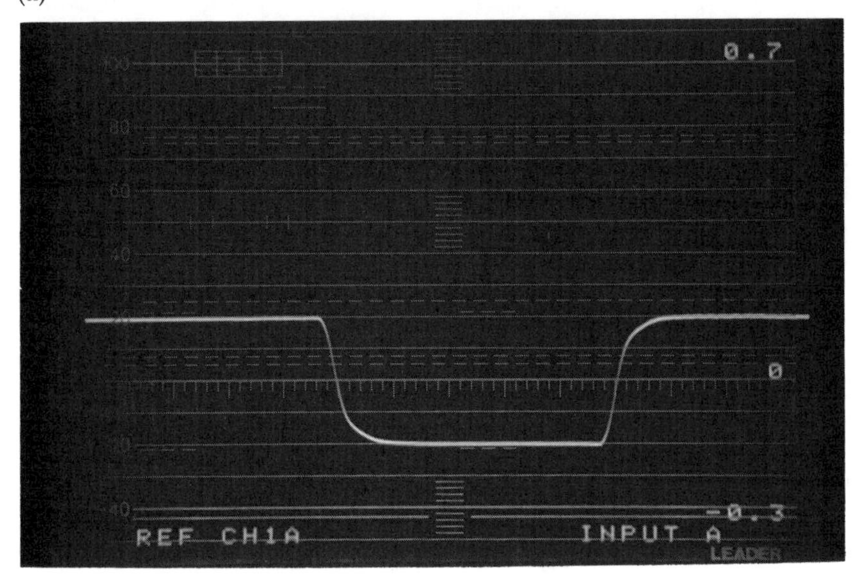

(b)

These are the line syncs of two cameras magnified horizontally five times. Photo (a) shows the waveform aligned horizontally to bring the sync leading edge to the centre and placed against a convenient time-scale point. (Note how the vertical shift has been adjusted upwards.) When the other camera is selected as in (b), its timing is seen to be late by 2 μs. Although there is no picture information on the waveforms, the signals are 'black and sync'. The test may be carried out just as easily in the presence of the pictures.

Figure 4.15 Timing the signal

counter that counts the lines down to the point to be observed, then releases the display for that line. This feature is particularly useful when studying aspects of the signal involving the four-field PAL sequence, explained in detail later on.

DC restoration is also provided so that the waveform will always have black level placed at the same point on the graticule. Subcarrier filters allow the separation of luminance from chrominance so that these may be looked at and measured more easily.

The scope is designed to be a versatile measurement and fault finding tool. Its input circuitry is designed for minimum effect on the systems and circuitry and the wide range of time and voltage scales available make it versatile. It is designed to accommodate a variety of waveforms, all shapes and amplitudes, audio, power and even DC.

Many scopes do offer additional video facilities but video is so complex with important parts of the waveform difficult to study with conventional displays, that the WFM has become a standard part of the video system.

In practice

Measuring the signal requires the same care as measuring anything else; the correct conditions and a steady eye. This means being in a comfortable position, with the WFM also comfortably placed. CRT-based WFM's produce light and are therefore quite critical to ambient light falling on the screen. The more recent use of LCD displays eliminate this problem for these are light reflectors; they require external light to be seen. Unless, of course, they have their own internal illumination, in which case ambient light may well require controlling.

In any case, in practical viewing situations, the presence of picture monitoring will necessitate ambient light control. Associated CRT waveform monitoring will therefore also be in the correct lighting conditions, Figure 4.16.

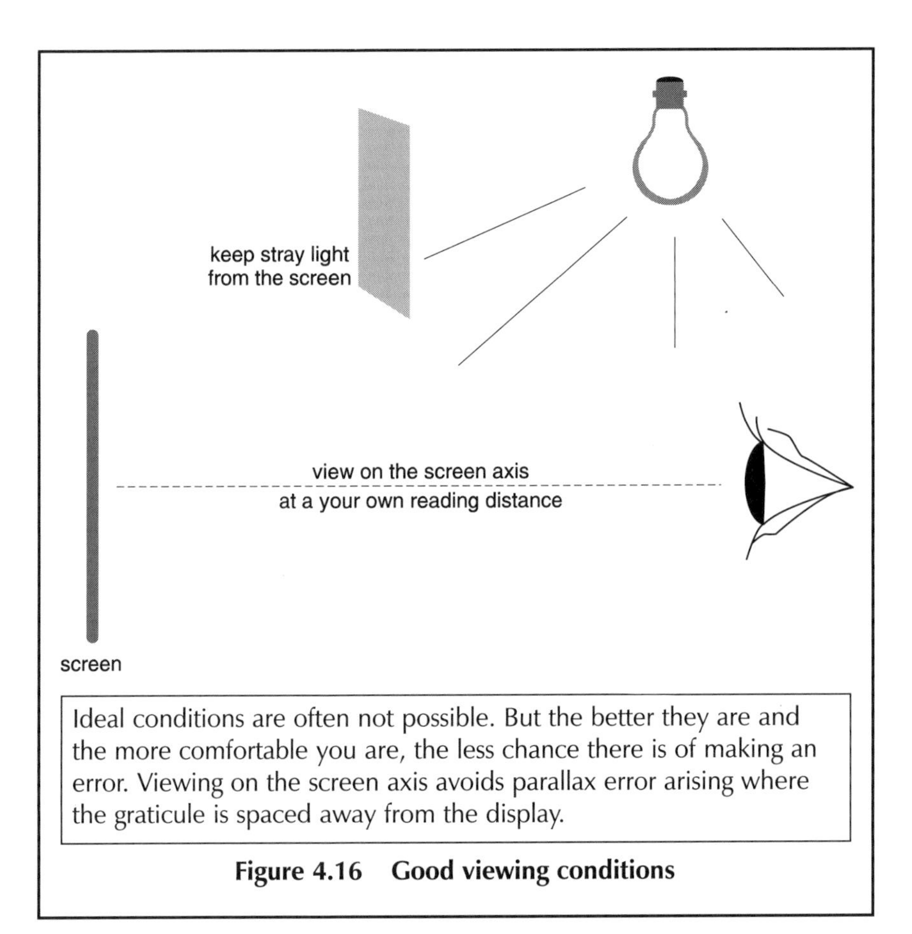

Figure 4.16 Good viewing conditions

CHAPTER 5

ADDING COLOUR

RGB and component

It is an established principle that any colour may be described in terms of red, green and blue (RGB). These are the **colour primaries** of additive colour; the distinction between adding coloured lights together and reflecting white light from coloured dyes is very important. Additive colour is that produced by the screen of a picture display, where red green and blue in equal amounts add to make white light.

We must also establish a few other principles before a full understanding of colour can be gained, starting with the picture. Where the picture display screen produces no light, the result is black (assuming ideal viewing conditions with no reflected ambient light). Where the screen produces the maximum possible of each colour, the output is white. Between these two limits is an infinite number of levels. At any level, where red, green and blue are equal, the colours add to make grey.

In practice, the black of one picture display may differ from that of another, depending on its design. The same applies to the maximum light output available. A consequence of this is that actual values of black and white are not absolute, they will vary from screen to screen. Standards must therefore be established before a working system is possible.

The choice of colours to use – which red or green or blue – was dictated by the availability of phosphors for the CRT's used in early colour television. That standard still applies but there has been a steady improvement in this field over the years. The one principle that does remain is that equal values of RGB produces grey and this has become established in the video standard. Whatever the picture source or processing applied, this standard always applies.

The values in Figure 5.2 may be set down as voltage levels, i.e. 0 V to 0.7 V, and these values would apply to inputs as well as outputs where the signals are sent over standard circuits.

The video standard states that when all the colour signals have the same level the result is grey. Therefore:

when R = B = G = 0 the result is black
when R = B = G = 100% the result is white
when xR + xB + xG = xY where xY is the equivalent luminance, or grey, value.

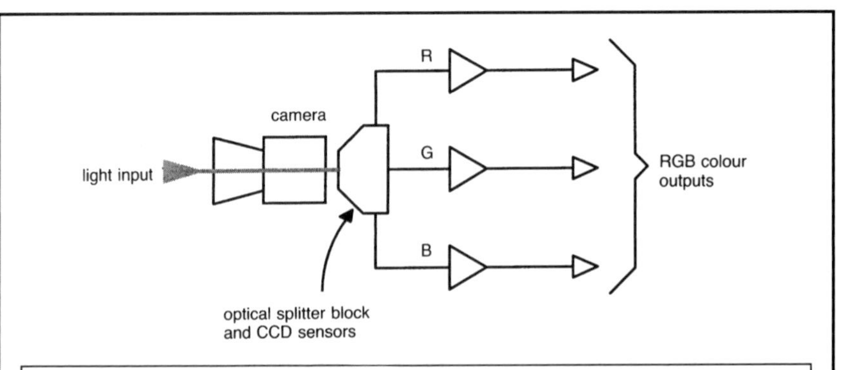

The light is split into red, green and blue by a dichroic optical block and converted by CCD's into RGB signals. The three-channel CCD camera is standard in the television industry. The RGB outputs are usually processed for the recording format, or to suit studio system requirements.

Figure 5.1 The three-colour video camera

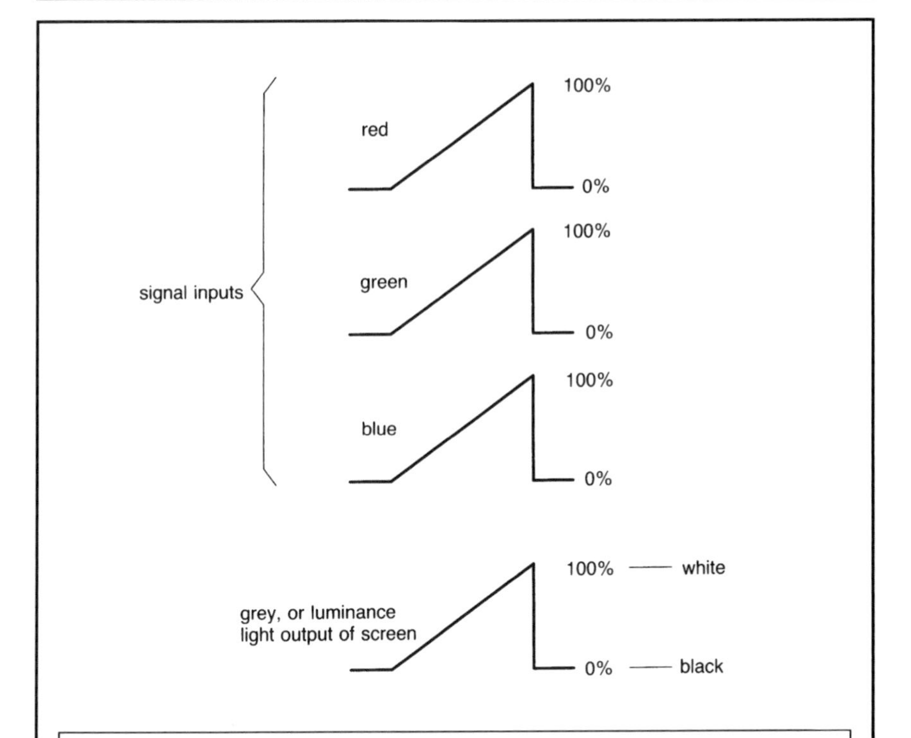

The ramp test signal is sent to the three colour channels of a picture display and produces continuous variation of grey tone.

Figure 5.2 Additive colour

Where the scene has only one colour present, say, red, neither green nor blue will be present. Other colours are a mixture of red, green and blue in various amounts. There are also the colour opposites to red, green and blue, which are cyan, magenta and yellow and are known as the **complimentary colours**. These appear when one of the primaries is not present:

R + B = Magenta. (Green is not present.)
R + G = Yellow. (Blue is not present.)
B + G = Cyan. (Red is not present.)

Colour pictures may be sent as RGB signals, but this is not a particularly efficient way to do so. The use of three circuits incurs three times the cost. All three circuits must be identical and all signals must experience the same processing. If either of these is not adhered to the principle of equal values would be violated and colour distortion will occur.

There are, however, some useful features in the way we see that enables some parameters to be modified and simplifications made. Our ability to see fine detail in colour is less than in black and white, colour can, therefore, have a lower specification with regard to how much detail information is sent. Lower resolution colour signals directly translate into a bandwidth saving. Our vision is more sensitive to differences of hue than of saturation and we are more aware of colour changes than we are of colour intensity. Another factor is that we are most sensitive to green, our eye's acuity peaks in the green part of the spectrum. Of all three colours, green most closely resembles the luminance signal.

Because of the way we perceive colour, benefits can be gained in the design of the system and the way pictures are transmitted.

As the black and white signal is so important in its own right, it is advantageous to separate it from the colour signal. Having done so, there was little point in sending the full RGB colours since each contains an element of luminance. Taking these factors together, the answer is to send luminance, plus the red and blue **colour difference** signals, that is the red and blue colours signals minus luminance. Green, because of its similarity to luminance, need not be sent for we only need two colour difference signals with luminance to reconstruct the complete colour picture.

The Y signal has the advantage in that it is the complete black and white signal and a usable picture is still available should the colour signals get lost, or if a black and white picture is all that is required. And the colour signals may be specified for colour resolution only, so trading off unnecessary signal capacity. This then is the basis of **component video** and is shown as:

Y luminance, all colours adding to make the black and white signal
R - Y red minus luminance
B - Y blue minus luminance

Green is recovered with a little mathematics worked out by the electronics in the picture display.

The requirement for all three signals to be maintained at the correct levels still applies. Should they vary then colour errors, both hue and saturation will arise.

Colour bars

The ramp test signal is a black and white signal and has limitations when applied to colour operation. Figure 5.2 shows how the three RGB channels carrying identical signals achieve a black and white output. The representation of colour values is best seen with the standard colour test signal, colour bars. This waveform is also useful to us in learning more about the colour system.

Figure 5.3 illustrates, with colour bars, the relationship between the colour signals and luminance. Subtracting Y from a colour signal produces a negative and positive going resultant, a symmetrical signal. The blue component is the greater with nearly 90% above **and** below the axis, a total swing of almost 180%. The red has a total swing of 140%. These are compared to the normalised maximums of 100% for the RGB and Y signals and far exceed the one-volt standard for video signals. At this point, it should be said that we are here considering approximations but these are perfectly adequate in this discussion about the principles of colour video.

The high amplitude values of the colour difference components shown in Figure 5.3 are wasteful; the colour information in actual pictures is usually quite small and does not justify this amount of signal space. A more equable solution is therefore to reduce the amplitudes of R – Y and B – Y to more realistic values. This would of course upset the principle of additive colour, however, by restoring the original values before the calculations are made by the picture display, the answer will turn out the same. The colour difference signals are therefore **weighted** to bring their amplitudes into line with standard values.

In doing this, we make yet another standard which must be agreed by both system designers at either end of a circuit. There is also an inherent practical problem. Reducing the value of a signal before sending means it must be amplified on its arrival. This will compromise the performance where noise is concerned because noise will be amplified as well but we have already said that we are less critical to some kinds of colour degradation; colour noise is one of these. Blue, the least critical in this regard, has maximum weighting; it was the larger of the two in the first instant and its weighted value comes down by half.

The weighted values of the component signals, as applied to colour bars, are listed in Table 5.1.

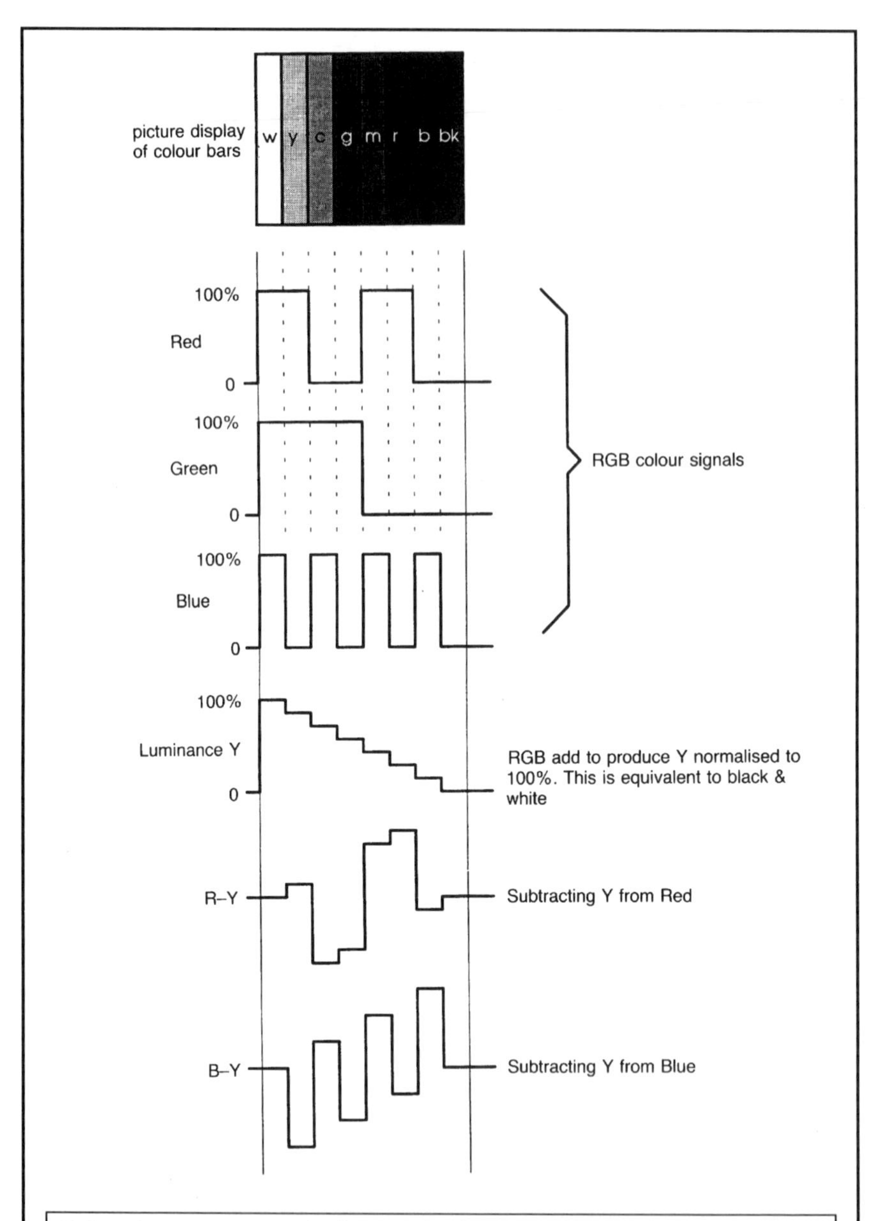

Colour bars represent the limits of colour and black and white signals. The three components, RGB, each have a luminance element. Subtracting Y from RGB produces colour only signals. Note how the R – Y and B – Y signals are very much higher amplitude than the individual RGB and Y signals, this results from subtracting the luminance value from a colour of lower value, producing a negative signal swing where the complimentary colours occur.

Figure 5.3 Colour bars and colour difference

Table 5.1 Weighted values of colour bars

	R – Y (%)	B – Y (%)
Yellow	10	– 44
Cyan	– 62	15
Green	– 52	– 29
Magenta	52	29
Red	62	– 15
Blue	– 10	44

The maximum values of component colour difference signals. The weighted red component has higher value allocated to it than blue, reflecting the greater importance attached to red. Note also how the colour compliments have equal but opposite values to their respective primaries.

The colour difference signals go below zero, to swing negative, producing the subtractive colour elements (for instance, minus blue is yellow). In consequence, it may be thought that such signals contravene the rule about the signal going below black level. A distinction has therefore to be drawn between standards for luminance and colour difference signals. As a consequence, only the luminance signal carries synchronising information, there is neither need nor benefit in more than one set of sync pulses going to the same destination.

The figures in Table 5.1 represent the maximum permitted, here shown as percentage. Because these values normalise to 100%, these are known as 100% Colour Bars.

Composite video

The separation of colour and luminance has become established in many areas of video operation. A number of variations of component working exist, adapted to different requirements and situations. But where pictures have to be sent any distance, the use of three circuits remains a serious disadvantage. For national TV transmission it is prohibitive. To bring all three into one signal in one cable is the next step and this is called composite video.

Essentially the black and white video waveform described in the previous chapters is adapted to accept the addition of colour. The bandwidth requirements remain essentially the same, allowing similar specification circuits to be used for colour working; in practice, equipment specification is marginally upgraded.

The Y signal is the same as the original black and white signal. Colour information is first placed into a high frequency sine wave carrier by a process called **modulation**, producing the **colour subcarrier**, or CSC, or simply C. The two, Y and C, are then added together to form the composite signal. The whole process is known as **encoding**.

Subcarrier is a fixed frequency of 4.43361875 MHz for PAL and 3.579545 MHz for NTSC. The choice of frequency is determined by various factors specific to the standards concerned. Carrier signals have been used in telecommunications for multiple speech transmission for a long time: the technique may be described as putting information in a protective envelope. When in this form, signals are less likely to interfere with each other, or be interfered with and distortions are more easily interpreted and put right.

The amount of colour information in actual pictures may be quite modest but its integrity is crucial for the eye's colour sensitivity is extremely acute. The protection afforded by the CSC technique is quite crucial to the accuracy of colour television.

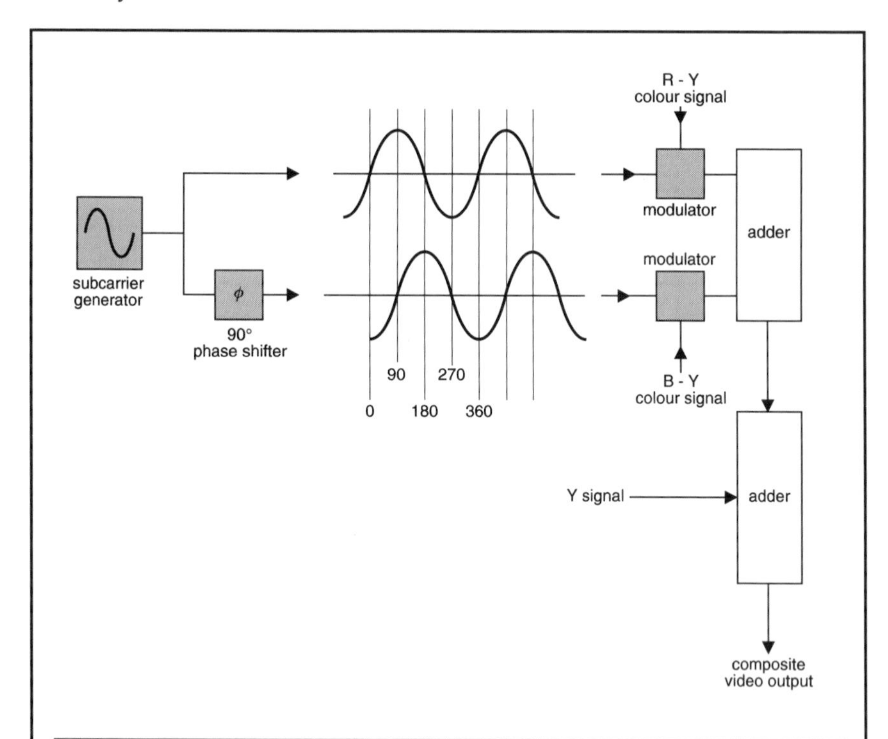

The subcarrier generator output is split for the red and blue information to be modulated onto it, the B – Y carrier is phase shifted so that it runs a quarter cycle behind the R – Y carrier, i.e. it lags by 90°. After modulation the carriers are added together. The resulting subcarrier (CSC) is a complex signal whose phase and amplitude varies with the colour information. The modulated CSC, or chrominance, is added to the luminance to make a composite video signal.

Figure 5.4 The basic principle of colour encoding

Figure 5.4 shows the principle of the colour encoder. The method of modulation used is called **suppressed carrier**. In practical terms this means that when the colour signal falls to zero, the carrier does likewise. Where the colour information is at a maximum, the subcarrier will also be at a maximum. By using the same subcarrier divided into two paths, one delayed by 90°, the colour difference signals are given separate but related identities in their modulated form. Each envelope will vary in amplitude with colour saturation and, upon being added together, produce a waveform of the subcarrier frequency but of varying phase and amplitude. Figure 5.3 and Table 5.1 indicate how complementary colours, yellow, cyan and magenta produce negative values of R – Y and B – Y. When modulated they take up the opposing phase relationship; yellow is 180° from blue, cyan is 180° from red and magenta is 180° from green.

The significance of this method of encoding the colour information is that:

1. Colour saturation is carried by the **amplitude of subcarrier**. Where there is no scene colour, subcarrier will be zero.
2. Hue is carried by the **phase of the subcarrier**.

The choice of suppressed carrier operation means that no transmission power and space is wasted by sending a carrier signal when not required, i.e. when no colour is present. If this were not the case, it would be like sending an envelope without a message inside.

These two points are fundamental to composite video. We can see that the composite signal carries all the colour and black and white information and it also has some very useful fail-safe features. Hue changes, to which the eye is so sensitive, is carried by the phase angle of the subcarrier and it is relatively easy to build in safeguards to protect this from becoming altered. The amplitude of the subcarrier is, however, more vulnerable but as this determines saturation, a less critical parameter, the effect on the viewer is minimised.

On arrival at the receiver, after what may have been a long and difficult journey, the signal may have suffered degradation and it is quite likely that the original waveform is less than perfect. The all-important information it carries must, however, still be recoverable. To do this, reliable decoding of the chrominance and unravelling the complexity of phase and amplitude variation, must be achieved. A practical picture contains all values of colour hue and saturation in varying degrees. To ensure the decoding is carried out accurately after the addition of transmission noise and disturbance, an accurate colour reference is sent as part of the composite signal.

The colour burst

To recover the colour information the decoder 'looks' for subcarrier. In areas of high colour it is easily found but it will be elusive in areas near grey, and none at all in where true grey exists. Chrominance is, by its very nature, a complex waveform and there is only one way to extract the colour information.

By sending a colour reference, we provide the means for the decoder to regenerate its own subcarrier of identical frequency and phase to that in the signal. A simple quartz oscillator is too inaccurate. Back in Chapter 3, we saw the requirements of video timing, how the accuracy of quartz clocks falls short of video timing requirements. By adding a short burst of subcarrier, called the **colour burst** to each line a permanent clock reference is sent along with the signal and the decoder is able to look for picture subcarrier simply because 'it knows where to do so'. See Figure 5.5.

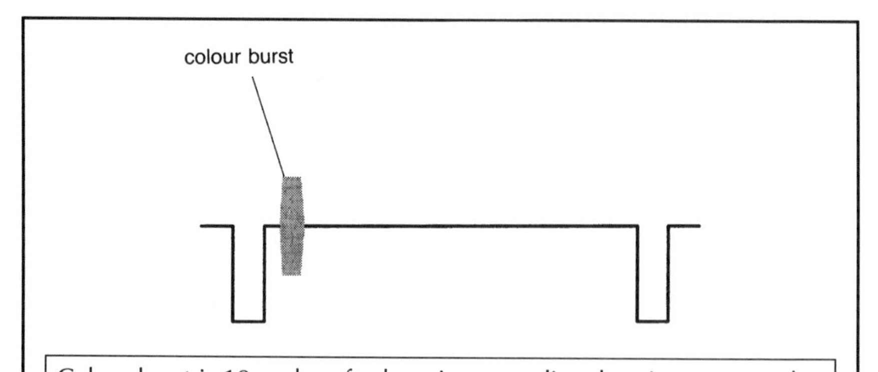

Colour burst is 10 cycles of subcarrier preceding the picture on each line, it is also present during most of the field period. It has a very precise and accurately maintained phase relationship to the rest of the signal. The colour burst amplitude is 0.3 volts peak-to-peak. (0.3 V_{pp}). The sync detection circuits in receivers, etc. are designed to ignore the negative excursion of subcarrier below black level. The waveform shown has no picture information and is therefore often called 'colour black', or 'black & burst'.

At the point of encoding, the colour burst must be accurately generated and maintained for it is this small part of the composite signal that provides both phase and amplitude information essential to the correct reproduction of colour. Colour burst serves to 'lock up' the decoder oscillator so that a perfect replica of the original subcarrier is accurately regenerated, perfectly stable in phase and amplitude.

Figure 5.5 Line waveform with colour burst

Figure 5.6 shows how the various components that make up composite video colour bars appear. Colour bars represent the maximum values permitted and so form an excellent test signal for the system. Current practice now states that 100% colour bars exceed the practical requirements of normal pictures. A less saturated version based on 75% colour saturation values is now standard for international circuits.

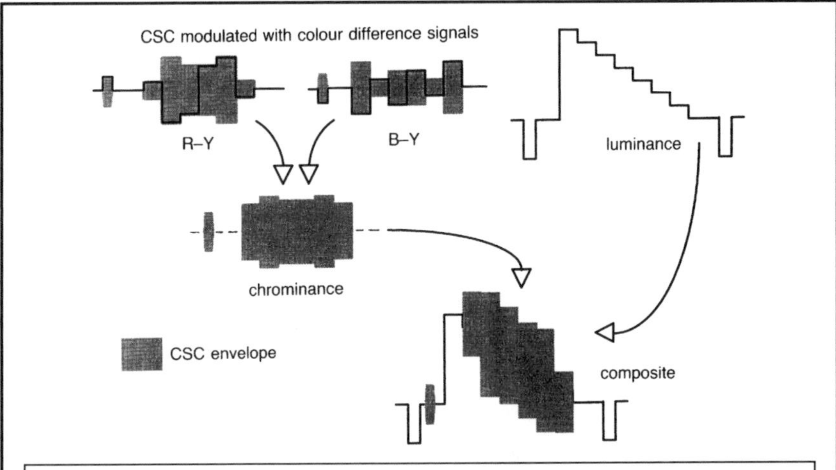

The colour difference signals are modulated onto the CSC, their original forms are overlaid for comparison. Note that where there is no colour, i.e. the white and black bars, there is no subcarrier. The colour burst is created with the colour difference signals and becomes part of the modulated envelope. The colour difference signals are added together to make the completed chrominance signal. Finally, chrominance and luminance are combined to form the composite signal.

Figure 5.6 Colour bars in composite signal form

Figure 5.7 is a very simplified overview of the decoding process, a practical design is a far more complex piece of equipment. For those desiring more a technical coverage, further reading is recommended. Other aspects of composite video will emerge as different operational procedures come under scrutiny and these will be dealt with as they arise.

Vectors

Colour information carried by the phase and amplitude of subcarrier may be translated into a vector, that is, its phase and amplitude corresponds to the **vector co-ordinates**.

In Figure 5.8, we can imagine, as observers, the sine wave streaming by, its voltage rising and falling at the frequency of the wave. We can also illustrate the action as a rotating vector that follows the voltage swing, not along a continuous axis, but about a fixed point. Take the argument one step further, place ourselves alongside the wave and travel at exactly the same speed. Now the wave appears stationary. If we could do the same with the vector, it too would be stationary. This is what the decoder does.

By having its own regenerated CSC locked to the incoming colour burst, the decoder is able to synchronise its observation of the signal to

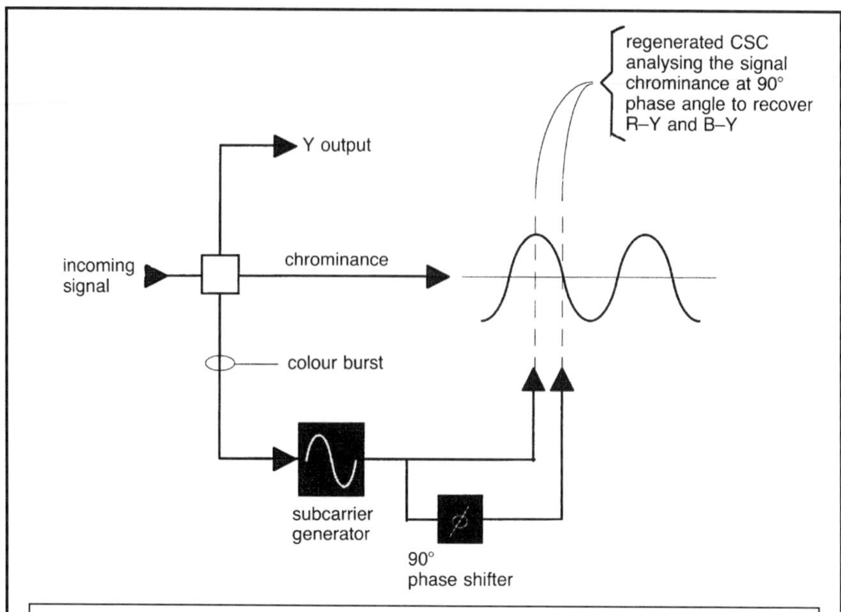

The incoming signal is split into subcarrier, Y, and colour burst. The colour burst is used to lock the decoder's oscillator to regenerate two outputs of CSC, one phase shifted by 90°. These 'look for' subcarrier in the signal, and where present, recover the colour difference signals.

Figure 5.7 Principle of colour decoding

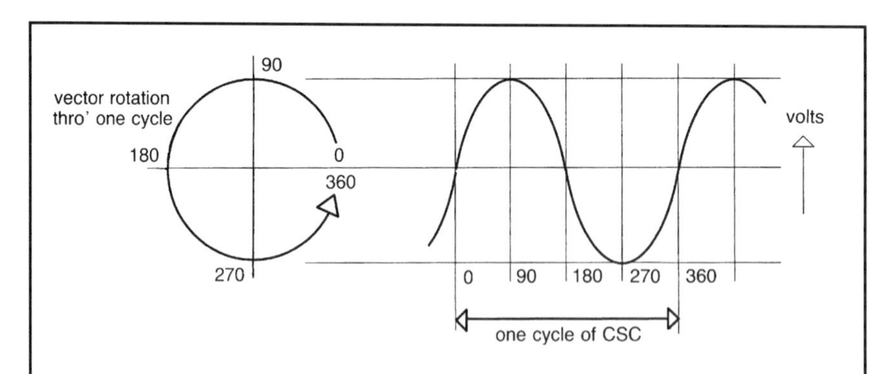

The vector constantly rotates to follow the sine wave's instantaneous voltage; as it passes through zero volts, the vector is at either 0°, 180°, or 360°. One complete rotation is 360°. By observing the vector at exactly the same point in the cycle, the vector will appear stationary. The length of the vector represents the sine wave amplitude at a specified point in time.

Figure 5.8 Vector rotation of a sine wave

measure against itself, the phase and amplitude of chrominance information. And so values of colour hue and saturation are determined.

Likewise, describing colours in vector form enables us to see them as stationary when our observation is synchronised to the signal chrominance. It's like the stage coach wheels in a Western; when they run at the same speed as the camera shutter, the spokes appear still. When the coach slows down they slip back, and vice versa. That's the relative phase shift-

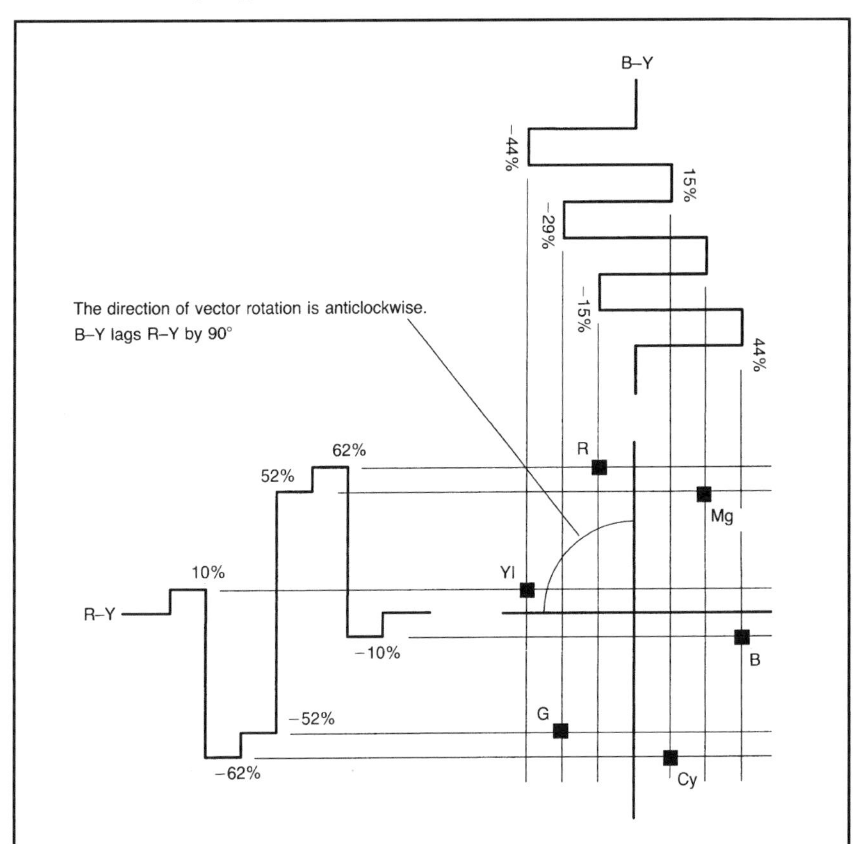

R – Y and B – Y are modulated onto the subcarrier, with B – Y delayed by 90°. Therefore, as shown here, R – Y amplitude is shown *vertically* and B – Y amplitude is shown *horizontally*. The signal colour values extend and meet at the vector co-ordinates that specify hue and saturation by vector angle and length. Angle is measured from the horizontal axis starting at '3 o'clock'. The vector for magenta on this diagram could therefore be specified as:

$(\sqrt{(52^2 + 29^2)}) = 60\%$ at, $\tan^{-1} \frac{52}{29} = 61°$

Figure 5.9 Colour difference vectors

ing about, an effect we call **strobing**, but in electronics it's known as **synchronous detection**. A good description of the colour decoding process.

Figure 5.9 uses the colour values of colour bars to show how a vector representation of colour is built up. As the colour varies along the line so the vector values change accordingly. Should the CSC shift phase through one complete cycle, that is 360°, it will return to its starting point. In doing so the vector will rotate through the whole colour spectrum, returning to its starting point. There are (theoretically, at least) an infinite number of possible colours in one 360° sweep, all having a specific vector value.

With the colour spectrum held in this way, the ability of the decoder to measure small phase angles is crucial to accurate colour reproduction. The colour burst gives a reference of phase every line against which the CSC phase variations on that line are compared, from which the colour difference components are reconstructed. To get some idea of what this accuracy must be, let us work out how long 360° of CSC represents.

The subcarrier frequency of NTSC is 3.579545 MHz, therefore, the length of one cycle is:

$$1/3.57945 \text{ MHz} = 0.2793725\ \mu\text{s}$$

and for PAL,

$$1/4.43361875 \text{ MHz} = 0.225549388\ \mu\text{s}$$

Considering that the whole colour spectrum is held within these time periods, only very small variations of timing will introduce a noticeable colour change. In fact, variations of greater than about 0.005 μs. are likely to be noticed. Earlier in Chapter 3, we saw how the video system relied on timing to synchronise picture displays to picture generators. A timing accuracy figure for black and white television of about 0.5 μs was considered good enough. The figure of 0.005 μs is one hundred times more accurate and timing to this accuracy requires a different approach.

If the standard requires an accuracy of 5°, then, $\frac{360}{5} = 72$, is the available hue resolution.

Rationalising the two standards for convenience and letting one cycle be 0.25 μs, we can see that getting 72 colours squeezed into this time-scale means working to within 3 or 4 nanoseconds.

With this timing specification required for colour phase in composite video, the measurement requirements are beyond the conventional WFM. But by separating the chrominance out and observing that alone, the problem becomes much more easily dealt with. Now colour can be most easily observed by using the values derived from the colour vector values of phase and amplitude.

Measuring timing values for synchronism now splits into luminance, which we have already discussed in Chapter 4, and chrominance.

NTSC and PAL

Further differences now begin to emerge between the NTSC system and its European derivative, PAL.

The eye offers other features that the composite signal is able to utilise in the interests of fine tuning the system. We are able to resolve fine colour detail better in the red/orange colours than in the blue/cyan. Bandwidths of the colour signals can therefore be adjusted to take advantage of this but to do so effectively, their axes should be realigned.

This means advancing the modulation axes by 33° from the original R – Y and B – Y axes to those approximating to orange/red and blue/cyan. These new axes are called I and Q respectively, the latter, carrying blue/cyan, having lower bandwidth requirements than the former. Trading bandwidth where it is not needed like this makes it available elsewhere in the signal. PAL does not make use of this feature and does not use I and Q modulation axes.

Although NTSC and PAL are technically incompatible, their basic processing principles are the same. Where the colour difference signals are shown in this text, it should be remembered that these are modulated on the I and Q axes in NTSC but remain R – Y and B – Y in PAL. And there's another twist: in PAL, the R Y and B Y signals are referred to as V and U.

Phase errors

The principal difference between PAL and NTSC is the additional protection PAL offers against the effects of phase variation. Let us now look at how this affects the picture.

Because hue is described by the phase or vector angle, and a 5° shift is considered the maximum permissible error, the whole system becomes extremely sensitive to timing. In standard coaxial video cable the signal travels at about two-thirds the velocity of light and over a typical studio length, say 50 m, the signal will take about 250 ns, (0.250 μs) to appear at the far end. This is a complete cycle of subcarrier, or the whole colour spectrum and so cable length is extremely critical where colour phase is concerned. A 5° shift corresponds to the delay in just under a metre of cable.

There are other considerations. Consider two colours, one of low luminance, the other, high, present on the same line – a typical situation of adjacent scene colours. As the signal passes through apparatus, the colours may experience different degrees of distortion, distortion that may alter their respective phase. Called **differential phase distortion**, there is nothing that NTSC can do about this. The colour burst has given the reference for that line; if the phase now alters during the line, it will be interpreted by the decoder as a change of hue.

PAL was developed to overcome these sorts of problems. It has an inbuilt protection to eliminate changes of hue arising from burst to chrominance phase errors. It is based on the principle of cancelling error by adding its opposite value. By switching the polarity of V (the R – Y axis) every line, any phase error apparent on one line will, in the next line, appear as the opposite error. To the eye the two will cancel each other.

To do this reversal the subcarrier for the V modulator passes through an inverter every other line. The V vector therefore swings through 180°. It is important to note how the colour burst is affected by this. It is a feature of both systems that colour burst is created as part of the colour difference signals and, is therefore modulated onto the CSC in exactly the same way as picture colour. In other words, the PAL burst and V chrominance switch polarity together and so the requirement for precise phase relationship of burst to chrominance is maintained.

A line is scanned by the display in the same place every other field, because of interlaced fields. When PAL switching is applied to the signal, alternate fields, although similar in line structure, have alternate V subcarrier polarity. Hue errors on successive lines will therefore cancel.

Perfection, however, is elusive. One can imagine that the eye will see the result of two opposing colours laid one over the other, as a reduction toward grey. In fact, desaturation does occur and if the phase error is great enough – right round by 180° – the colour will disappear entirely. But, for practical purposes, desaturation is to be preferred to colour change.

So how does NTSC deal with the problems of phase and timing? Simply by using good engineering design and maintenance standards, although, it must be said that when NTSC exhibits phase errors, they are most unpleasant. An indicator of how a practical system can fall short of the ideal.

In fact, since these systems were designed, technology has developed and improved so much that the protection afforded by PAL could be considered redundant. Well, this is an academic question; whilst a standard exists it has to be used correctly. But there is no doubt that PAL can, and does unfortunately, encourage design ‘corner cutting’, particularly in some low-end systems.

The vectorscope

The vectorscope is really a precision composite video decoder with a vector display instead of a picture. It is similar in many ways to the waveform monitor but whereas the latter views linearly the vectorscope is rotary. The trace actually travels outwards from the centre to mark the colour vector positions. There is a rotary shift to position the display over the graticule and magnification to enlarge the trace so that low saturation colours can be observed – those that are very close to grey (or black or white). These will lie at or near the screen centre which is zero colour.

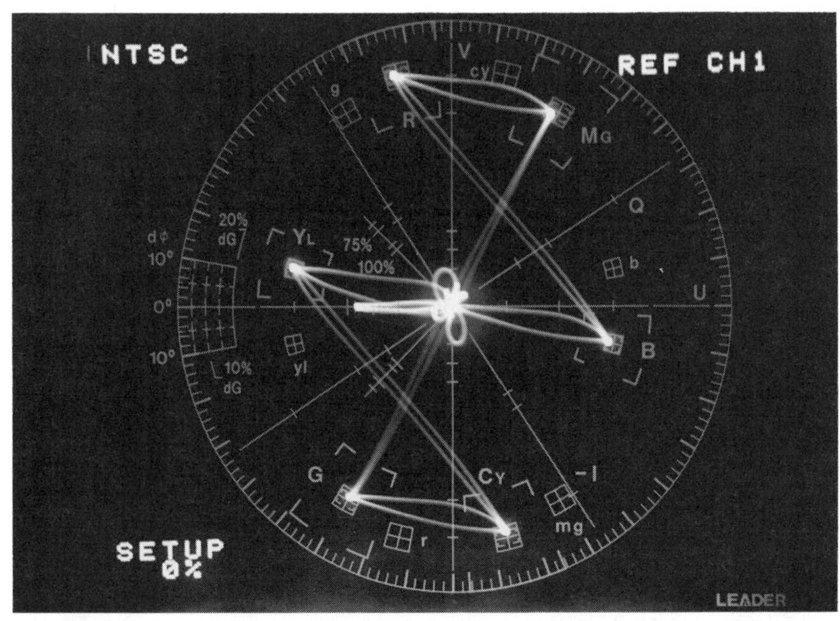

(a)

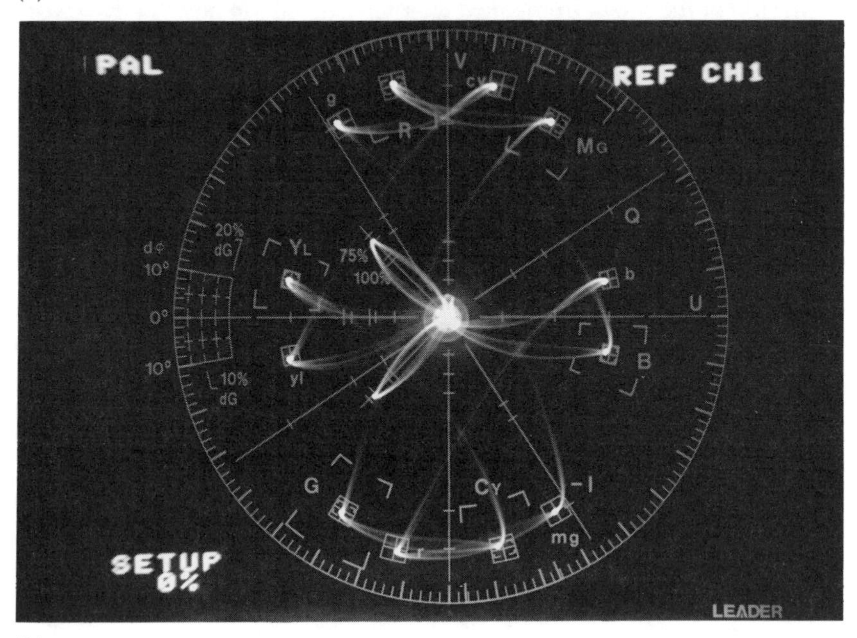

(b)

Vectors only show colour information; luminance values always appear as zero, i.e. the display centre. NTSC is shown in (a) and PAL in (b). Notice the effect of 180° subcarrier V phase change on alternate lines in PAL.

Figure 5.10 Vector display of colour bars

In addition to variable magnification, there is also provided 75% and 100% gain positions. These correspond to 75% and 100% colour bars, with the graticule markings, or boxes, for the fully saturated colours remaining the same. Both sets of bars will fall into these boxes when the correct gain is selected. The colour burst, however has two marks for amplitude as the burst remains the same at 0.3 V_{pp}, for both 75% and 100% colour bars and will therefore change vector length when the gain changes from 75% to 100%.

Like a WFM there may be more than one input and it may be possible to display more than one signal for comparison. An external reference input will be provided which can be a standard composite signal or black and burst; it may in some cases be derived from one of the inputs. The vectorscope will ignore everything but the colour burst when taking its synchronising reference from the signal.

Such is the basic hardware similarity between vectorscopes and WFM's that many manufacturers combine them into single units. The screen, CRT or other, is the same and switchable graticules are easy to provide. The combination offers lower overall cost and there is no serious disadvantage in this arrangement.

Vectorscopes and waveform monitors are complementary – neither can do the job of the other. In the next chapter we will see how the two work together.

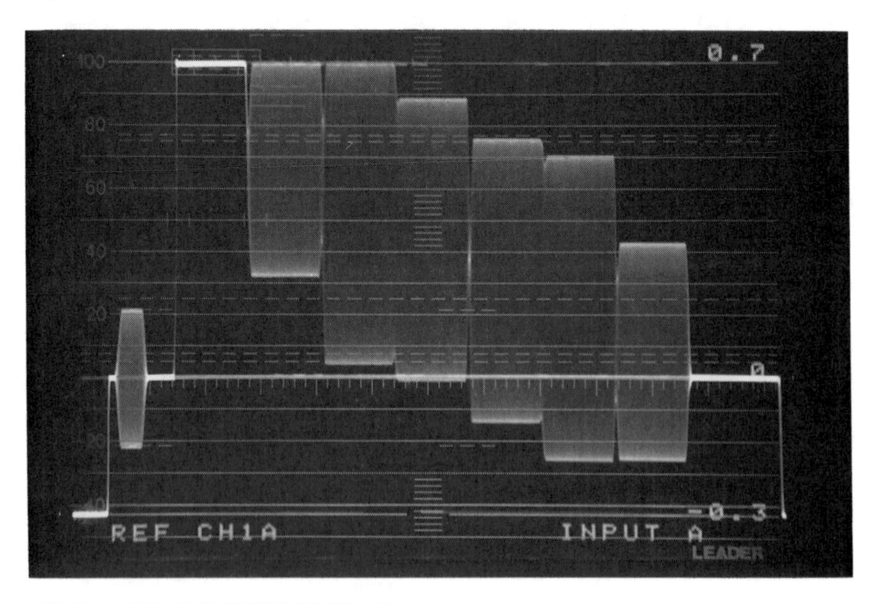

75% colour bars. The WFM is unable to display any information about colour apart from saturation which shows as the amplitude of chrominance.

Figure 5.11 WFM display of colour bars

CHAPTER 6

THE SINGLE CAMERA

We have discussed at length the background to video signals and measurement. Now is the time to look at practical situations. First there are the picture sources and then, how these link into systems. The most complex picture source is the camera because unlike any other source, its output is photographic and is therefore unpredictable as regards tonal range and colour.

But before we lose ourselves in the excitement of camera check-out, we must deal with yet another feature of the video system if, in the end, our check-out is to have any relevance at all.

Terminations

Chapter 1 laid the groundwork to understanding circuits and terminations and described the principles of sending and receiving video signals.

The standard video connector is the BNC, a bayonet locking connector that preserves the circuit characteristic right through from cable to equipment. Proper termination is necessary to ensure the transmission of a video signal along a cable of any length, with minimum loss and distortion. As part of this design standard, the source must be correctly loaded, or terminated. Only when this is done will the signal appear at its desti-

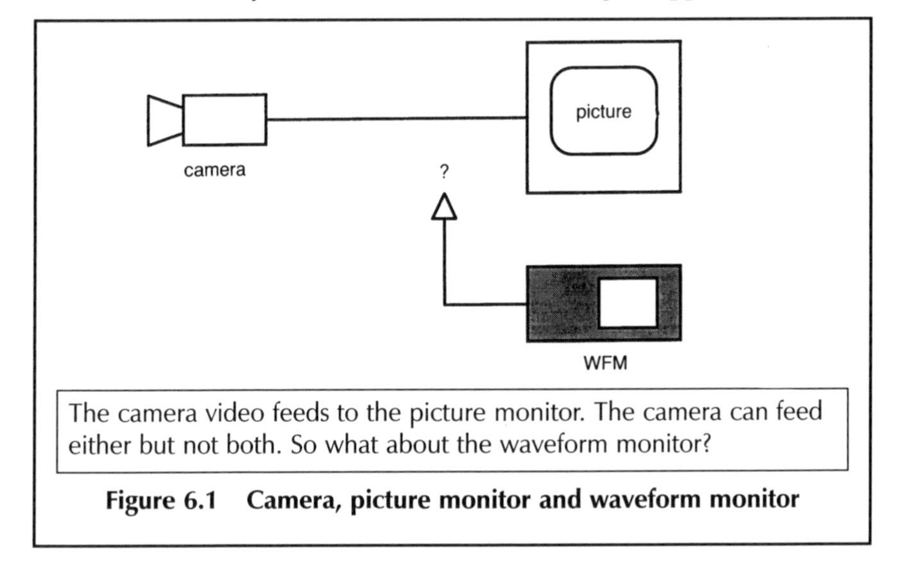

The camera video feeds to the picture monitor. The camera can feed either but not both. So what about the waveform monitor?

Figure 6.1 Camera, picture monitor and waveform monitor

nation at the correct level and with minimum distortion.

Figure 6.1 illustrates a common problem: one source, one destination. Video is a complex waveform that requires correctly designed circuits, to comply with the standard of one source for one destination – all sources, whether cameras, video tape players, or test signal generators, may only feed one destination. Where sources have more than one output, each output will be isolated from the others, so becoming a separate source in its own right. For this reason it is common practice to have more than one output.

Many lightweight cameras may only have a single output. So how is the WFM in Figure 6.1 accommodated?

There are two ways to deal with this. Ideally, the camera should feed a distribution amplifier (DA) which is simply a device to split a single source to more than one destination. A DA has one input and more than one output. But there is a more common alternative.

The inputs of most professional equipment, such as picture and waveform monitors are 'looped' inputs. A looped input does not present a termination and does not draw power from the source, it merely 'observes' the signal. The signal has still to be terminated somewhere along its route, so, where the signal feeds two items of equipment, the last in the line provides the termination.

The standard termination load resistance is 75 ohms (75 Ω) and draws the required amount of power from the source. 75 Ω is characteristic of the coaxial cable used for video transmission.

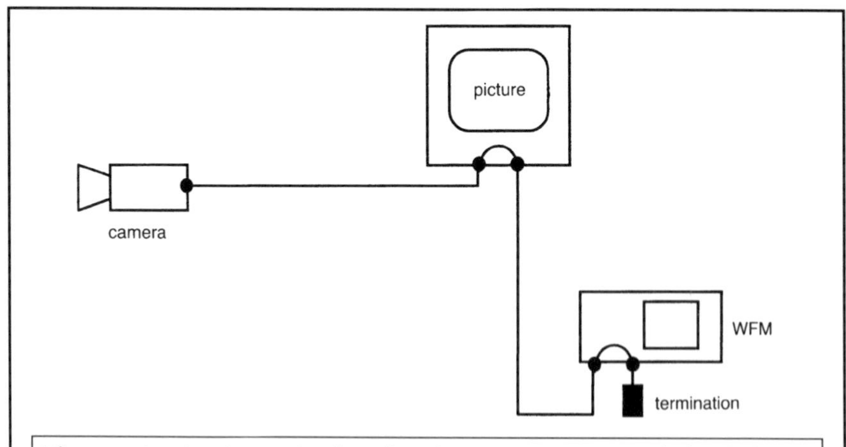

The termination is required so that the camera is correctly loaded, only then will it produce the correct signal level. Here, neither picture monitor nor waveform monitor load the signal as each input connection has a pair of sockets. The sockets are joined internally so that either may be used as the input whilst the other feeds the signal on and, in this case, a termination is fitted at the final socket.

Figure 6.2 Using looped inputs

(a)

(b)

When camera and monitor are properly set up and the signal correctly terminated, the picture will appear as in (a). When unterminated, the result is as in (b). By reducing monitor contrast, the effect may be allieviated but *the fault will remain*.

Figure 6.3 What incorrect terminations look like on picture

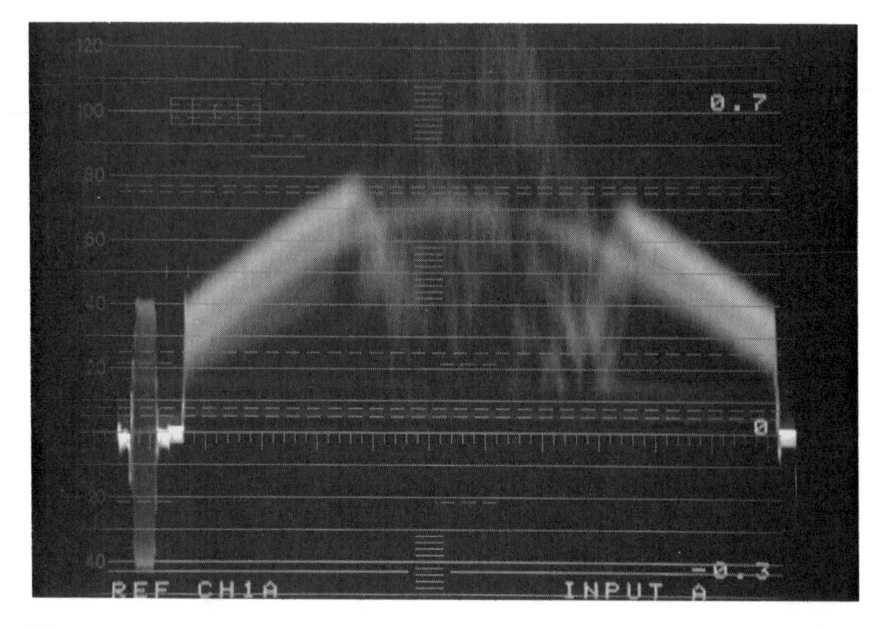

(a)

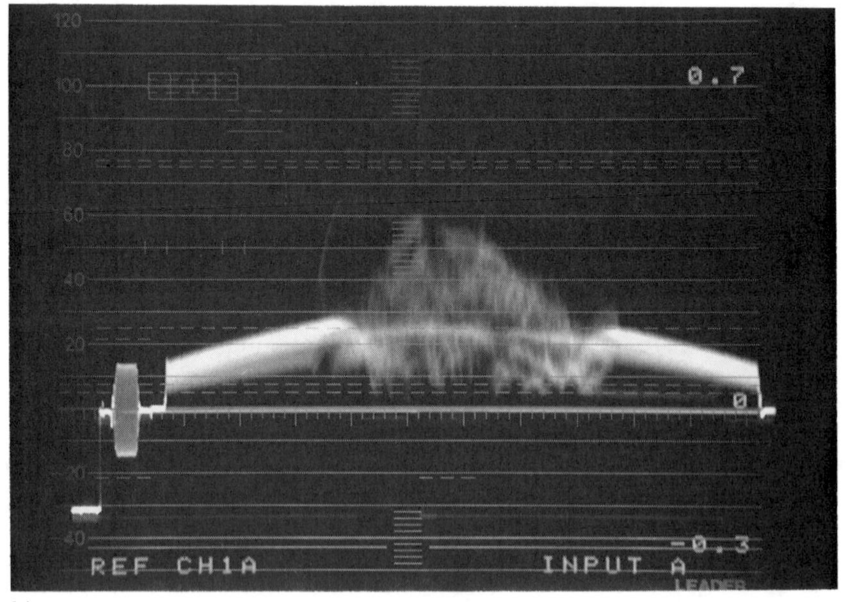

(b)

The waveform of Figure 6.3: (a) the signal is unterminated and the amplitude is twice that of a terminated signal, resulting in signal levels well beyond the permitted maximum of 100% or 0.7 V. The signal is double terminated in (b) and the amplitude is reduced by a third.

Figure 6.4 Incorrect terminations

Looping through is rarely provided on domestic or similar equipment for in these circumstances it is more liable to misuse thereby giving rise to termination errors. It is usual to provide a fixed termination on such equipment. However, equipment with fixed terminations are not entirely free from abuse: the use of standard cables and connections would, ideally, make the system foolproof, were it not for the availability of various kinds of video cable adapters.

The T-piece is a most commonly misused adapter; it allows two cables to be connected to one socket allowing additional destinations to be placed on one source. Understanding of all this is not helped by the various socket configurations provided on equipment. The most common methods are described here:

Single socket with fixed termination. The input terminates the signal.
Single socket, unterminated. Requires a T-piece either to feed on or to be terminated. Now less common.
Single socket with switched termination. Requires T-piece to feed on. The switch must be in the correct position to terminate or not.
Pair of looped sockets. Most common on broadcast equipment. The input can go to either socket, a termination, or terminated cable; must be placed on the other socket.
Pair of looped sockets with switched termination. No T-piece required. The most versatile. But beware – the termination may be automatically switched when a cable is connected. Check the labelling carefully.

It is well worth consulting manufacturers' handbooks of the equipment concerned if in doubt about connections and terminations. If all this 'termination talk' seems a bit heavy going, Figure 6.3 and Figure 6.4 show what happens if you get it wrong.

It is unfortunate that modern picture monitors are very forgiving for in the situations shown in Figures 6.3 and 6.4, the monitor controls could easily have made good the termination errors to make the **picture** seem OK. Figure 6.5 illustrates one of the most common problems.

Camera colour bars

Before starting camera checks it is well worth spending a little time looking through the camera manual. The language may be rather unhelpful or too technical, but there should be basic information about connections and menu displays that will be important. It is not the intention here to detail every step; in any case, cameras differ one to another. The aim is to acquaint you with a procedure that applies to most systems and, indeed, much of what is said will apply to other picture sources as well.

On switching on the camera, a few basic functions can be quickly checked. The presence of a viewfinder picture and what, if any, information there is in the viewfinder, such as battery condition, light level, etc.

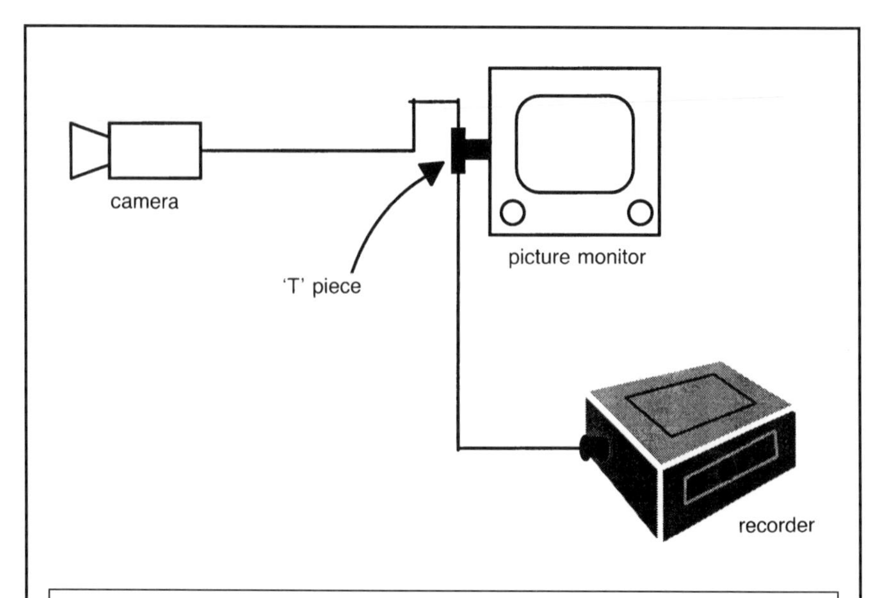

Both monitor and recorder have single input sockets and standard terminations. Together they double terminate the camera's output. But because the operator re-adjusts the monitor contrast until the picture 'looks right', the error goes undetected. In fact, the recorded picture is only two-thirds its correct level. The contrast control which the operator adjusted is a **gain** control, that is, it adjusts the signal level, but only of the picture.

Figure 6.5 Double termination

A broadcast, or equivalent camera has a 'colour bars' function. Whenever a recording is made, or a signal sent by any other means, it is convention for colour bars to precede the pictures. They will be accepted as representing the maximum values of all the colour signal parameters at the destination and will be used to set up receiving apparatus, video tape players and so on. Your pictures must therefore conform to this standard and, generally speaking, camera design ensures that this is so.

Select 'Colour Bars' on the camera. As the standard test signal, colour bars may be relied upon as conforming to the video standard but we will confirm that this is so. Looking at colour bars on a picture monitor will prove very little about how accurate the bars are. A WFM is essential to check out colour bars, and a vectorscope as well if available.

In the previous chapter, we learned what colour bars are and how they are generated. Connect the WFM to the camera output – not forgetting the termination. You will find the WFM handbook helpful here if the unit's back panel labelling isn't. Connect up a picture monitor as well, loop monitor and WFM to properly terminate. It is all too easy to disregard the picture when carrying out engineering checks. Learn to relate

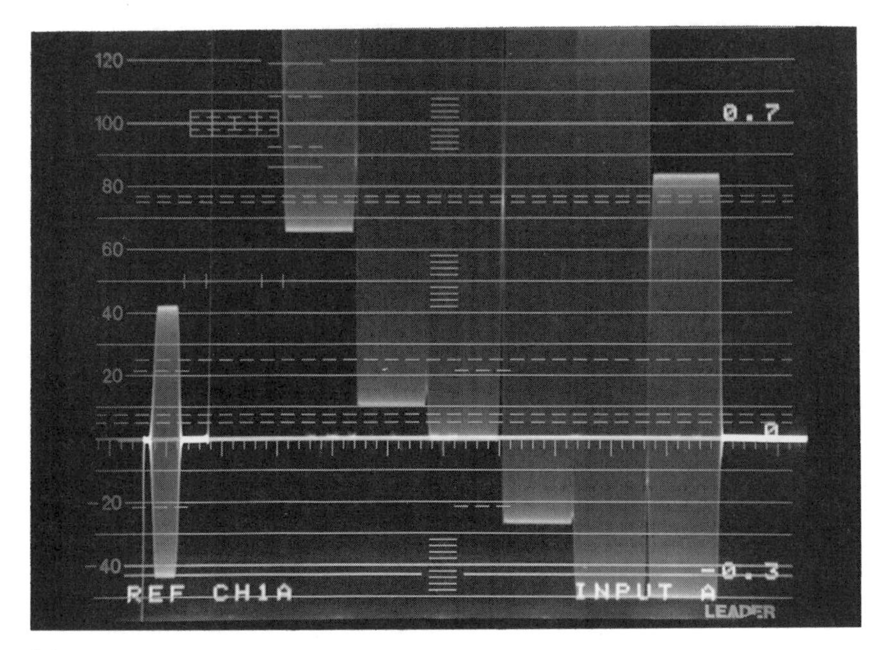

(a)

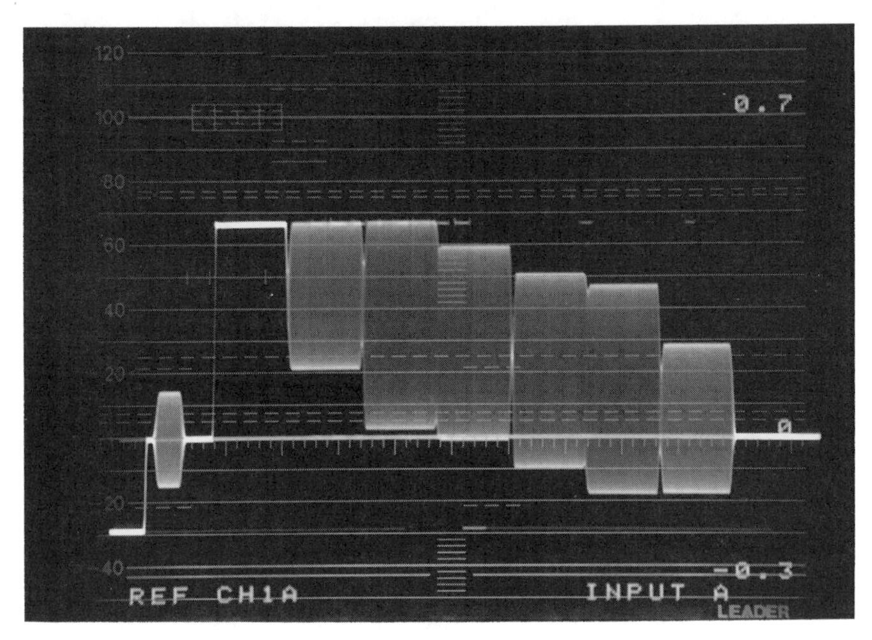

(b)

Colour bars (a) unterminated and (b) double terminated.

Figure 6.6 Incorrect terminations

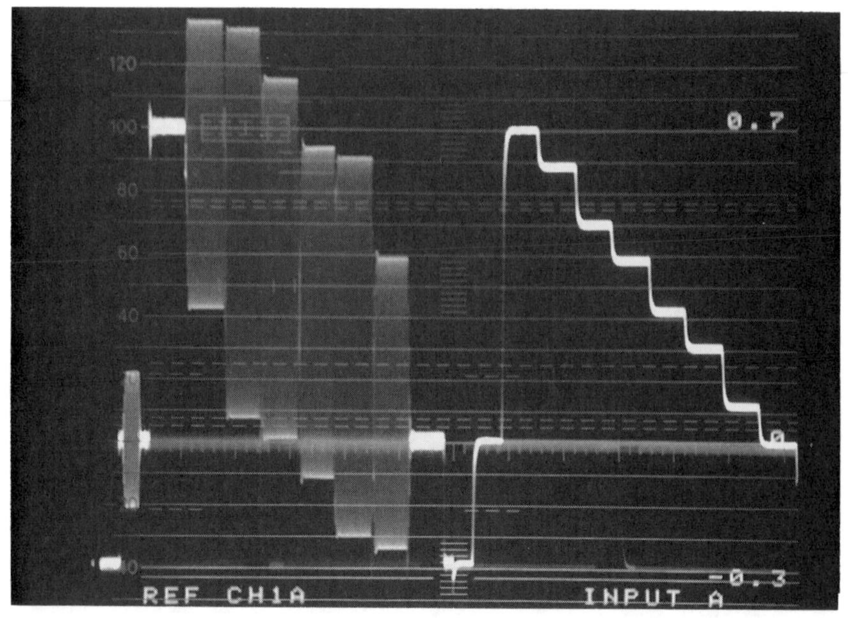

(a)

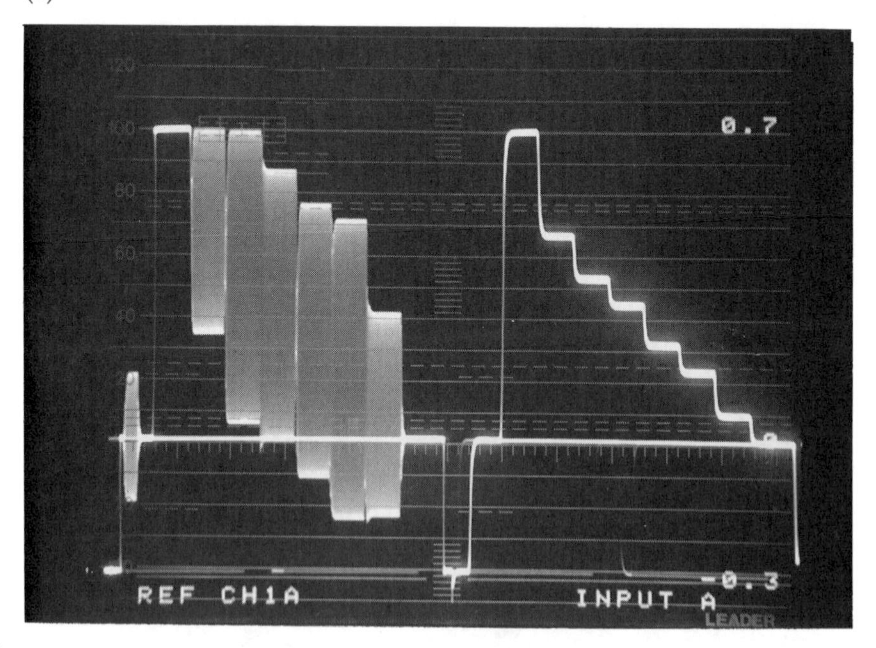

(b)

100% colour bars are shown in (a) and 75% colour bars in (b). Both displays show full composite and luminance only using the 'parade' function. Note how the luminance values differ between the 100% and 75% versions.

Figure 6.7 100% and 75% colour bars

the waveforms to the pictures they represent, so very important if a full and rounded knowledge of video is to be acquired.

With all connections in place, there should be a display of colour bars, waveform and picture, Figure 6.7. Note that these will normally be 75% colour bars.

Check the level of the white bar is at 100% or 0.7 V above black level. The syncs are 0.3 V below black level and the colour burst is 0.3 V_{pp}. Once more, NTSC and PAL throw up another subtle difference; this relates to ratio of picture to sync amplitudes. NTSC sync and burst are not quite 0.3 V. The Appendix has the details.

All this amounts to a check of the composite output of the camera whilst the recorder will usually be taking the signal in component form, but we have proved a large chunk of the signal path. To complete the test, record a half-minute of colour bars and look at these played back. Some older camcorders may not playback all that well, providing just basic picture monitoring: refer to the manual if in doubt. If not good enough, prove the recording by using a full specification playback machine.

If any deviation is noted from the correct colour bar form, the situation should be dealt with by a qualified engineer. It is unlikely that adjustment will be available without dismantling the camera. What degree of error is acceptable depends entirely on the circumstances of the camera's use. Do not rule out flaws in your own measuring methods; checking these must form part of any investigation into signal errors. Advice should be sought if you are in any doubt.

Colour bars are generated to high standards and problems are not common and where these exist, they are usually incorrect luminance and chrominance levels. If the luminance level exceeds 100% it is liable to be removed at a later stage in the production process. If lower, then this represents a waste of available signal space. If the white bar measures above 95% the camera may be considered acceptable, but anything in excess of 100% is not. Acceptability will be a decision based on many factors. Before you condemn, make sure your measurements are accurate and you understand the implications of what you do.

Saturation error will show by whether the green bar reaches down to black level. If there is a gap between black level and the lower edge of the green bar envelope, the saturation is too low; where it goes below black, it is too high. It is correct when it coincides with black level. The check is valid for 100% and EBU 75% colour bars.

When bars are selected on the camera, the circuitry will often close the lens down. That is, the lens iris will close to stop light entering, a condition known as 'capping up'. Now, only picture black is produced by the camera. Switching off the colour bars restores the picture, but take the lens manual, i.e. remove the camera's control over the lens iris, and close the lens again. A mark on the lens iris ring denotes where this point is, usually by a 'C'.

The picture on the monitor should have gone black; if it hasn't, the picture is not set up correctly. In the Appendix picture set-up will be described but for the moment let us assume our picture is OK. The waveform monitor should now show black, sync and colour burst, referred to as 'colour black' or 'black and burst'. Note there should be a detectable level change from video black in the front and back porches and either side of the colour burst, to picture black. This is the **camera pedestal**.

Pedestal should raise picture black by about 3% above video black, although the value may be user set. No signal, apart from subcarrier chrominance in the composite signal, is permitted to go below video black at 0%. Any attempt at doing so will be removed by the camera's black clippers. Pedestal is a degree of protection against any scene black being inadvertently clipped.

The setting of pedestal is an operational choice. It is adjustable, either through a manual control or via a menu function. If a menu is provided, the values shown may not relate to signal percentage levels, again the camera manual should explain the relationship between the two.

At this point another PAL/NTSC difference emerges. NTSC signals are often generated with about 5 to 5½% pedestal, or 'set-up' as it is more usually known in NTSC jargon. The PAL system leaves pedestal for the operator to decide, allowing its use as a photographic tool, just as in a studio environment, as will be seen later.

Returning to our camera, *the waveform will not necessarily reveal* if pedestal has been set below video black by the camera's previous user. There is only one way to check and that is to adjust. Watch the waveform and alter pedestal one way and then the other until you see it rise above the video black as seen in the porches. Adjusting up will clearly show, when adjusted down it will stop at 0%. Now open the lens and repeat the test. Observe both WFM and picture and see how first black, then dark grey are clipped at 0%. Flicking back to colour bars for a moment and adjusting pedestal again shows the test signal is unaffected by pedestal. You can work out the significance of this for yourself.

The option to set pedestal wherever one wants does present the risk of getting it wrong. Whilst discussing the faults arising from setting it too low, we must not overlook what effects too high a setting will have. Increasing pedestal raises black to grey softening contrast and offers a very useful pictorial device if that is what is wanted. But it does reduce the available picture contrast range. Hence, a setting of 3% is a good all-round setting for most work. Many photographers prefer to alter contrast by optical filtration, leaving pedestal alone and so avoiding the risk of getting it wrong.

It is to be hoped, however, that better understanding will bring about more willingness to use this valuable control for entirely pictorial effect.

Measuring 3% is often not that easy. It is a very small increment to see on a WFM; some instruments may have the means to measure such small values, but many may not. In a later chapter there is a totally different method that offers predictability and accuracy. But for the moment let us

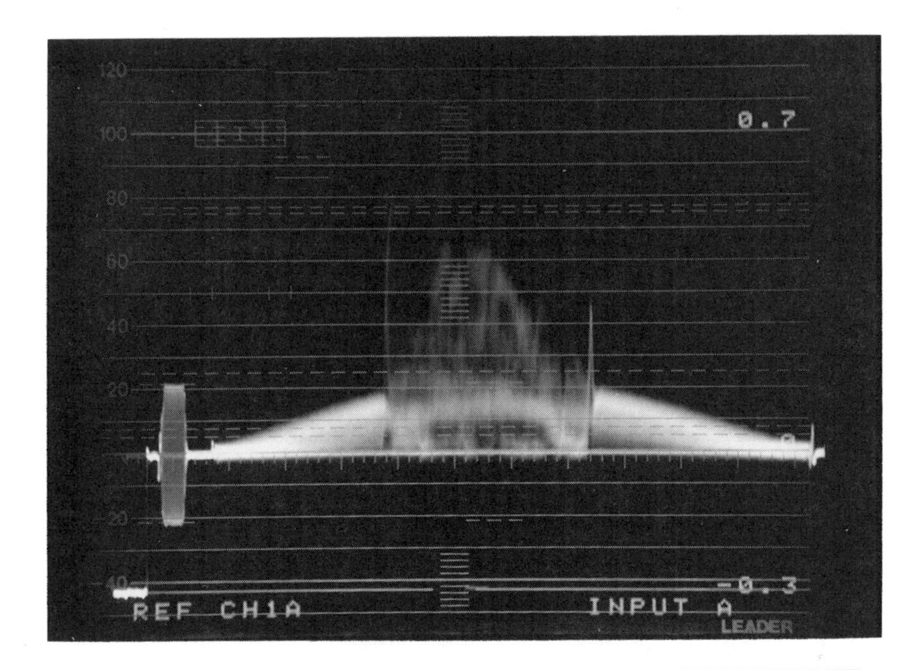

The waveform indicates that the picture is 'black crushing'. The tell-tale signs are black specula and larger areas of darker tones all going down to black level. These clipped picture elements are merged into black level and are not recoverable.

Figure 6.8 Pedestal set too low

see what 3% means. It corresponds to 21 mV (0.021 V) and estimating 20 mV is probably the best one can aim for. The actual value is less important than achieving a degree of pedestal around that figure. It is only when we work with more than one camera that precision is required in the matching of one camera pedestal to another. But that's yet another topic for later.

The Appendix has further information on contrast values that will be useful in comparing percentage to voltage levels.

Having checked out colour bars and ascertained the pedestal level, we have completed a major part of the camera check-out. Many more parameters remain in the camera; some will not be possible to deal with in a simple procedure but as the process unfolds, various observations can be made that will provide a very useful measure of confidence in the camera's performance.

The picture

When using a picture monitor it is imperative that you are able to see it properly. The monitor must be in subdued light, although the scene or set-up chart for the camera, must be well lit. If you cannot see the picture reliably there is little point in having a picture monitor for it will mislead and confuse. We are now going to study the picture and measure with the WFM and relate the two together and to do so requires a check of monitor set-up.

The main controls on the monitor are **brightness** and **contrast**. A simple test to prove how important it is to get these correct is to increase camera pedestal and then reduce monitor brightness and see how one cancels the other. It is this sort of confusion that has caused problems over pedestal setting. A similar interaction exists between camera iris and contrast.

There is no reliable answer to this without the introduction of monitor set-up equipment. If such is available, make use of it. Alternatively, many modern monitors address the problem by having preset positions for the controls but even here, there is no allowance for variation in viewing conditions. Nor can we exclude the possibility that preset control may not be correct. Such is the precision required in monitor set-up. For this reason we must for the moment, divert our attention from the camera to how we view its picture. There is little point in continuing any form of camera assessment if the picture we use is unreliable. The procedure described will concentrate on setting brightness and contrast.

Setting brightness requires that when black is sent, black will be seen – a contradiction in itself. It is 'the seeing of black' that causes the difficulty. Capping the camera means camera pedestal, which is picture black appearing on the screen. Raise the monitor brightness until this becomes grey, then reduce until black. Having discussed the careful setting of pedestal, such a method of setting monitor brightness is far too inaccurate and will not permit proper assessment of the picture.

Colour bars offer nothing in this regard; the signal has no elements able to provide even the roughest guide to brightness setting. What is required is a method that takes into account the individuality of picture monitors and, most important, the viewing conditions. Such is only possible with a picture set-up test signal. Picture line-up generators are available that produce a signal based on a black ground with patches at various levels from 5% to 100%. Brightness is set so that black and 5% are just resolved, whilst the light output from the 100% white can be measured and contrast adjusted as per the monitor manual.

The Vical method is designed for the single camera and monitor operation. Its principle differs from all other monitor set methods by inserting its test signal into the camera signal. Vical is connected between camera and monitor and by selecting it to Set-Up, video black, a variable level calibrated grey and 100% white, are inserted into the camera output and displayed on the monitor as part of the camera signal. With the camera capped, brightness may now be very accurately set by observing the difference between the video black from Vical and camera pedestal. Vical is described in more detail in the chapter describing alternative measurement techniques. Whichever way is chosen to set up the picture, make sure it is reliable.

Before we look at the camera's picture, there are a couple of quick observations to be done with the lens capped. When no light enters the camera, there should be no output. Ideals are rarely achieved and the camera invariably produces some spurious signal where there should be nothing, particularly older models. Raise the monitor brightness until the camera black appears as dark grey. Remember, this is camera pedestal; also remember what black is – it's no colour, with red, green and blue at zero. On our monitor what we see may not be quite so perfect. Look carefully for changes in brightness and colour, but you must keep the screen dark to do so.

If after a little careful study you do see imperfections, possibly shading in colour or luminance, look at the WFM. Can you relate the two displays? Look for tiny specks, usually white; these are where the CCD sensor chip has 'drop out', the failure of individual pixels. Another is colour shading, a black that is not truly black. Older tube cameras suffer badly with shading, modern cameras less so. A more common condition is colour shift overall due to a black balance error in the colour processing circuits. Such errors are not easy to see on the WFM; the eye is very critical to black errors and the picture will often reveal what the waveform does not.

This simple test highlights the need to understand what pictures actually tell us.

Whilst many could argue that camera deficiencies like this are really only academic. After all, who operates a camera with the lens closed? Faults often show symptoms, observation is part of good engineering practice. It takes practice and experience to see problems before they become serious.

Now open the camera lens.

Point the camera at a suitable chart. It could be a proper **television greyscale**, a pair of Kodak Gray Cards, a picture or whatever there is to hand. Obviously it is well worth investing in a greyscale, one that conforms to the reflectance standards used in television, as comprehensive measurements will then be possible. Greyscales differ in the reflectances of the steps as, unfortunately, not all manufacturers conform to a single standard. It may be considered useful therefore to have to hand Kodak Gray Cards. These are based on a photographic standard of 18% reflectance, which is a reliable international standard and are available world-wide. The accompanying white card is approximately 90% reflectance. Although only constituting two steps Kodak Gray Cards provide a known value of related light input for the camera. For this reason the following procedure uses grey cards.

(a)

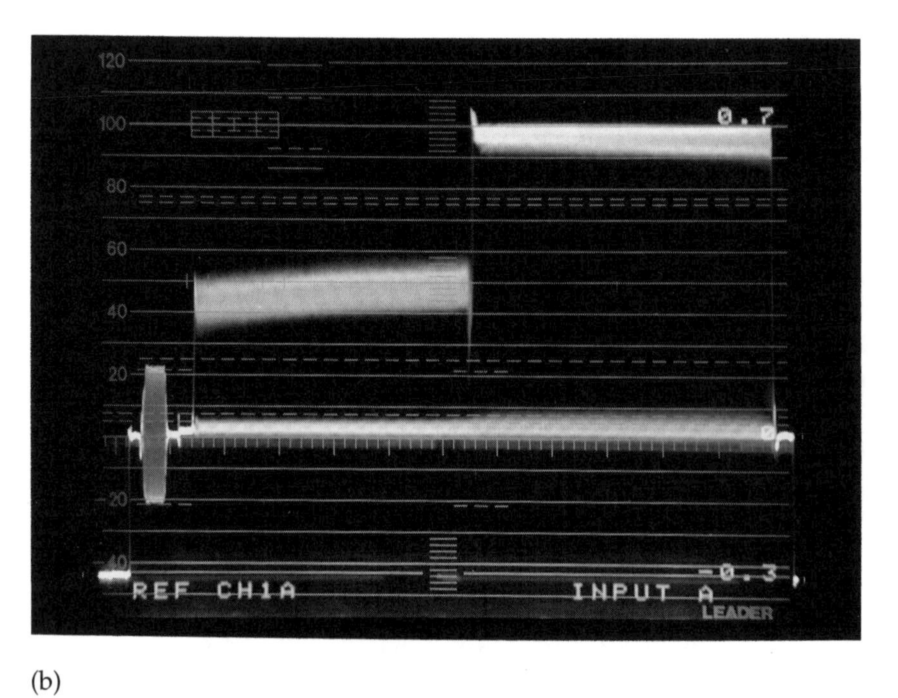

(b)

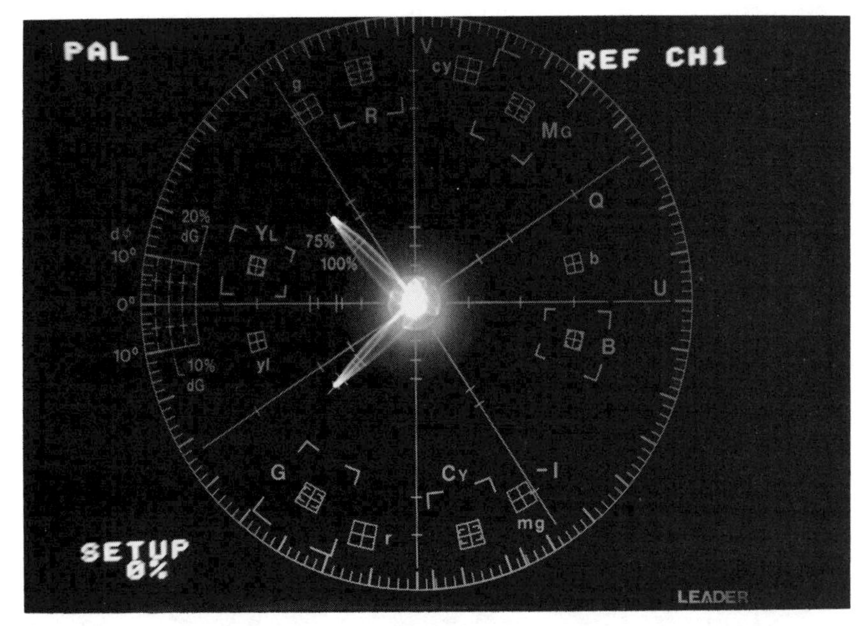

(c)

The cards (a), white on the left, grey on the right and, at the bottom, black for reference. The waveform is shown in (b). The camera is exposed to bring the white card to 100%; the grey card level reads 40%. Note that the waveform level is not consistent because of the variation in light level across the cards. The picture does not reveal this so easily – an instance of where the WFM tells more than the picture. The vectorscope display in (c) is at the centre showing that no colour is present.

Figure 6.9 Grey and white cards

Light the cards evenly. Use a proper studio light on a stand; there is no need for the lamp power to exceed 1 kW, but always keep the light off the monitors, particularly the picture. Expose the camera by opening the lens iris. Study the WFM and recognise the various picture elements. The white card will be the highest level of the waveform; adjust the iris to bring this to 100%.

The grey card reading of near to 50% is correct for the **gamma** of a broadcast standard camera. Not all cameras will give a reading of 50%: some may read lower, say, 40 or 45%, indicating a higher gamma. The Appendix explains the background to gamma and what it means in practice.

Increase the exposure until the white card is seen to limit on the waveform and will appear as shown in Figure 6.9. The level at which this takes place may vary camera to camera but for a broadcast camera should be

100%. Many cameras exceed 100% in the desire to increase available contrast range. Although the recorder back (if a camcorder) will be designed to accept such a level, the standard still remains at 100% maximum and any excess will be removed before sending through the transmission chain. The operator must decide whether this is acceptable to the operation. Adjustment of clip level is an engineering job but will be described in the camera workshop manual. See Figure 6.10.

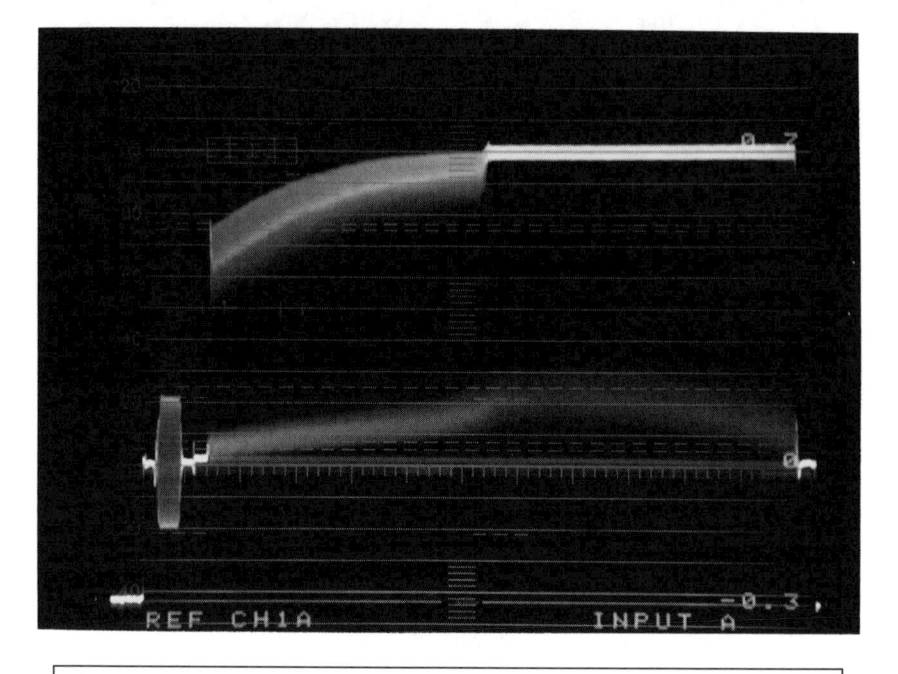

The camera is over-exposed and the white card is clipped back to 100%. Compare this to Figure 6.9(b).

Figure 6.10 Peak white clipping

The use of cameras with higher than 100% maximum output is becoming commonplace. Whilst very creditable in making greater scene contrast available, the limit of 100% will apply if the signal passes through international circuits. In considering the effect this has on a camcorder operating in the field we must know if the material is likely to fall into the broadcast category. If so there are two options: set the camera to limit at 100% by adjusting its peak white clippers, or, reduce the signal overall level during editing. The latter is usually beyond the control of the photographer and the result unpredictable as to how well it will be carried out – if indeed it is attempted at all.

Clipper adjustment is not to be carried out lightly. There are three clippers, one for each colour, and if they do not match, peak white will have a colour cast. Adjusting one alters the overall luminance value and getting all three to make 100% equally takes patience.

Contrast control is possible in various ways to alleviate the problem of excessive scene contrast. The signal is only allowed to rise slowly as it approaches 100%, causing compression of the higher tones as it does so. Peak white clipping will still take place but there its effect will be less pronounced and less obvious. Figure 6.11 shows how contrast control operates.

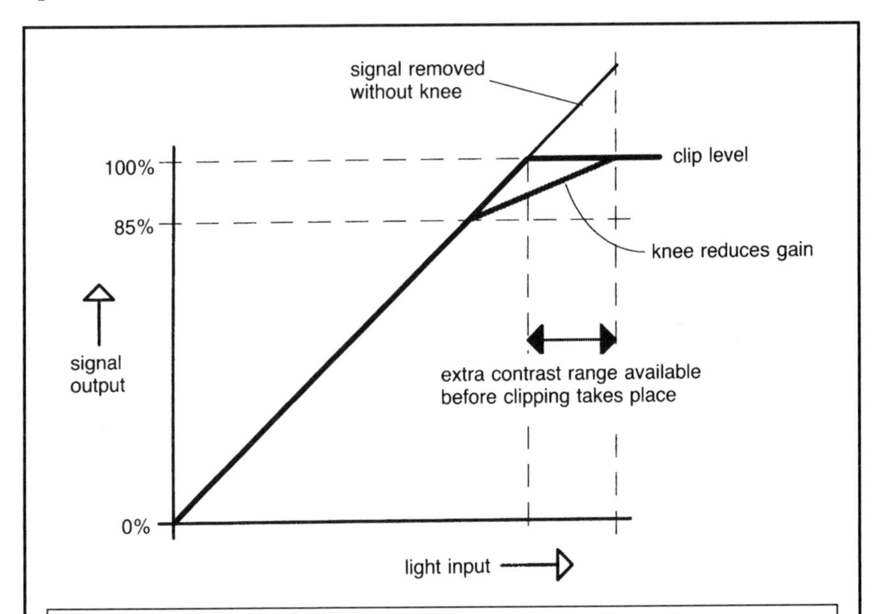

How contrast control works is shown in a graph of signal output against light input. Sometimes called a 'knee', it reduces the amount of signal increase above 85%. Effectively, this means reducing gain, or sensitivity, causing compression above this point. An increase of scene contrast is accommodated whilst still complying with the 100% limit. Dynamic contrast control automatically adjusts the compression depending upon the scene.

Figure 6.11 Contrast control by using 'knee'

Contrast control may be seen operating when the camera points at a white card, it may not permit full exposure of the fully framed white to 100%. The effect may be even more pronounced with dynamic contrast control (DCC). DCC continuously adjusts as scene content changes: it may reveal itself as allowing white specula to reach 100% but holding down broader areas of white to below that limit. Therefore, when conducting the tests described here, the facility is best switched off.

Dynamic contrast control methods are generally much more sophisticated than a static arrangement but the option to switch off may still be useful in certain instances. The camera's manual should explain the specific methods used.

Colour balance

Readjust for correct exposure with the white card at 100% (not clipped). White has no colour, nor has grey and the waveforms should show no evidence of subcarrier. Subcarrier will be present where there is colour and if the waveform shows subcarrier, as a thickening of the trace, the camera white balance is incorrect. White balance is the camera's way of making greys and whites neutral. But it has to be told what the kind of light is used to illuminate the scene. Daylight is more blue than tungsten artificial light; the Appendix describes what the differences are. Cameras may have both these light values as preset white balance options. They may be variously called 'daylight' or 'tungsten', or 5500°K and 3000°K.

Auto white balance is where the camera measures internally the three colours and adjusts them until they are equal. It assumes that the light entering is representative of scene grey and will adjust the gains of RGB to make grey. The camera must not be over-exposed and any value of grey from 50% to 100% is usually acceptable. Auto white balance may be offered as a continuously updating option, as the camera moves, so the scene is constantly assessed for colour balance. Presetting white balance is to be preferred for it will avoid constant and, possibly unsatisfactory, colour changes.

After deciding from the manual what is appropriate, operate the white balance and note the result. Also note that the monitor picture may not reveal the presence of colour at these higher tonal values; the eye is too forgiving of colour balance errors and so they may pass unnoticed.

Figure 6.12 is typical of colour error arising from the incorrect lighting condition for the camera setting. Here, the camera is set up for daylight and the greyscale is illuminated by tungsten light.

Colour errors occur when the colour channels are not identical. Gain differences make one colour dominant overall; lift or pedestal differences cause black to have a colour cast. A camera picture may show a colour change from dark tones to highlights. The grey card may have a colour cast that is not evident in the white card. It will take a practised eye to observe the effect unless it is quite marked. If distinct, the WFM will reveal presence of subcarrier on the grey card but not on the white. The effect is due to poor tracking of the colour channels; equal values of RGB to make grey has not been accomplished at all levels.

A camera that will carry out colour balance at black and white should produce true black and white, but a tracking error will be revealed when the midtones are not true grey. The fault reveals itself in use because skin tones appear false. The eye is particularly sensitive to facial colouring – subtleties that are beyond the ability of the WFM to reveal. The effect is often attributed to gamma differences in the colour channels and it is only the eye's considerable sensitivity to colour variations that makes the problem so acute. A 1% gain or level difference in colour channel matching will give a noticeable colour shift.

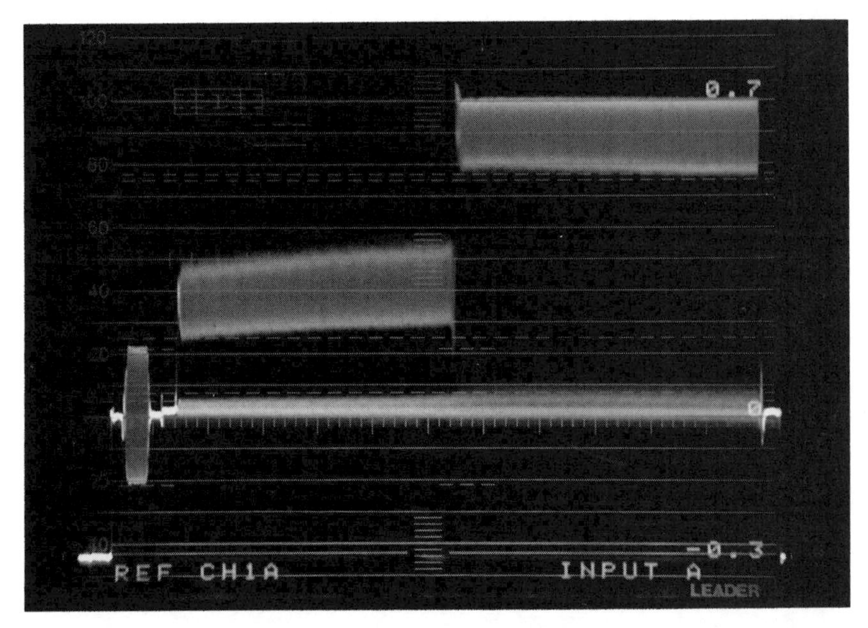

(a)

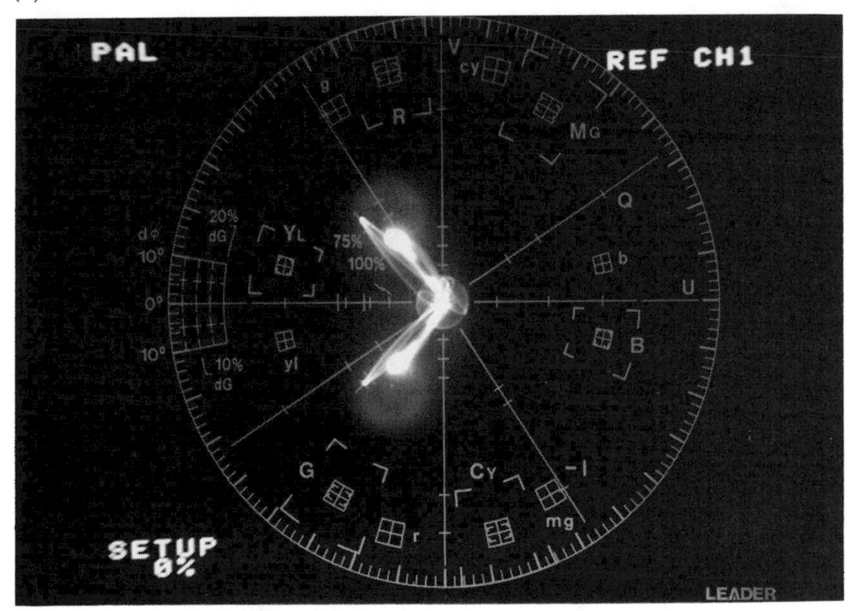

(b)

The white balance is incorrect and the grey cards are not reproduced as grey and chrominance is present. It appears on the WFM (a) as subcarrier and causes a thickening of the trace. The vectorscope (b) shows the colour as a shift away from the centre of the display on the red/yellow axis. Compare with Figure 6.9.

Figure 6.12 Camera white balance error

All this may appear rather indeterminate when looking at your picture; you may not always be certain if a colour error exists or not and we can be forgiven for questioning the value of the exercise. But this is a question of experience in looking at pictures. The camera is able to balance grey to very fine limits; our ability to observe these may be less predictable. In the case of a single camera, if the errors are not clearly obvious, the camera's performance is probably acceptable. It is where more than one camera is in use that such small errors immediately become obvious to the eye. When the eye sees a lie, there is the risk that this will interfere with the story being told. We must teach ourselves to see the error before the viewer does.

The reader will by now be quite aware that, although we have a WFM, possibly a vectorscope as well, a great deal of the discussion has centred around the picture. This is as it should be. Fine colour differences like these are not what the WFM is there to tell us about. Our picture is the important part – after all, who watches waveforms at home? To learn the waveform and how it relates to the picture, that's the important thing to do. It is only the picture that will tell you whether the error is acceptable or not.

Sensitivity and exposure

Cameras have different sensitivities: it is an indication of how effective a camera performs in low light. Sensitivity is linked to exposure. The correct exposure for one camera is identical to that for another; the white card should expose to 100% in each case. It is the setting of lens iris to do this that will reveal the differences in sensitivity from camera to camera.

To measure sensitivity requires knowing the level of light reaching the camera lens. The lens iris, or aperture, is calibrated in f-stops. Changing the iris by one stop alters the light transmission by a factor of 2 (see the Appendix for more information). Sensitivity is calibrated in lens f-stop for a given light level. The value of this is to be able to compare one camera to another. Although a camera's ability to function in low light is desirable, it does not preclude the photographic skill of the operator in using light creatively to produce fine pictures. That's required whatever the camera.

An indication of correct exposure is provided in the viewfinder and known as **zebra**. The name originates from the form of the diagonal moving pattern that it used. Zebra appears when picture elements reach a specific level; measuring this is part of the check-out procedure. Increase camera exposure, and the white card rises in level and zebra pattern will appear in those areas exceeding a predetermined value. Further increase eventually raises the grey card to the same level and that too becomes subject to the zebra pattern. Reducing exposure causes the pattern to disappear. Note the level on the waveform at which zebra appears, making sure that you identify the correct point by relating picture to waveform. Measure the level on the waveform.

The point at which zebra comes in is adjustable and a personal choice. It is also influenced by the actual method utilised in any particular camera. A good guide figure is 90% for this will alert you to levels approaching 100%.

Camera sensitivity is variable in steps, sometimes called **gain** and occasionally **sensitivity**. The calibration is in **decibels**, abbreviated to dB's, or may be expressed as a factor. Table 6.1 shows how gain and sensitivity are related. Sensitivity, in stops of exposure, is a direct relation to the rating of photographic film.

Table 6.1 Gain and sensitivity

dB's	–6	0	6	9	12	18
Gain (factor)	×0.5	×1.0	×2.0	×3.0	×4.0	×8.0
Sensitivity (f-stops)	–1	0	+1	+1½	+2	+3

Modern CCD cameras have variable shutters, they may be mechanical or electronic. They stem from the need to shutter the light from the sensor during its read-out time and are now adopted as an operational feature. The camera shutter sets the exposure time. The original tube cameras exposed at 1/25 s per frame but for CCD's this is now more usually 1/50 or 1/60 s. The exposure at other shutter timings are:

Shutter exposure, s	1/50	1/125	1/250	1/500	1/1000
Relative exposure (f-stops)	0	–1	–2	–3	–4

The presence of 1/60 s shutter is so similar to 1/50 as to make little practical difference to exposure.

It is worth considering the form of graticule scale used on the WFM. We have described signal level in two ways: picture levels are more meaningful in percentage terms but below black this is less so. Voltage scales starting a sync level, making black 0.3 V are not convenient for picture measurement. If the principle of relating picture and waveform is accepted then it is reasonable to use a scale of percentage. A scale of percentage engraved on one side of the screen and voltage on the other would be ideal but for the conflict of two sets of graticule lines. Switchable scales are the ideal answer.

Adjustments and simple checks

Modern video cameras are becoming more like film cameras, particularly from the way they are perceived. Film is a medium of limited adjustment compared to video. Video has traditionally been very 'tweakable', an indicator of its roots in the valve era where every parameter was liable to drift and adjustment was part of the operation. Now, video has turned these adjustables into features, providing numerous options that may be user set. These are not remedial adjustments but pictorial alteration. The types of adjustment are numerous but the common ones are:

Detail is the control of sharpness;
Gamma is modification to the camera's linearity (explained in the Appendix);
Colorimetry is how the camera sees colour.

Detail

Detail is not resolution: that is determined by the number of sensor pixels and the system bandwidth. It does however, have a similar effect by modifying the way transitions from black to white and vice versa, are made. To make these transitions, or edges, appear sharper, an enhancement is added. For example, putting a black line around a bright object will make it stand out more. Likewise, a dark object is given a light edging.

The concept is based on the fact that the small TV picture needs this kind of edge enhancement. It may also appear on some picture monitors and TV sets, a point to be aware of when viewing detail on the camera. As the system has improved, higher resolution and contrast range, traditional detail, with its hard black outlines, has developed into a much more subtle technique.

Setting the detail parameters is not difficult but does require a good picture monitor and a little care. Refer to the camera manual for specific instructions and watch out for the terms used. Detail is now accepted as a term by most manufacturers but within the designation there are subsidiary adjustments with a range of terms applied to them.

The sort of control availability offered is making detail level dependent; for instance, at, or near black, there is less need for effect. More sophisticated is the reduction of enhancement in facial tones but retaining sharpness in the eyes.

Setting up detail must be carried out on-picture but care must be taken in observation for the effect should not be excessive. It is very easy to overdo detail; the static situation of looking at a chart, or picture, or whatever, is not actual operations. Experiment by all means but be methodical: note the control positions – less easy if these are tiny presets; the menu system has much to offer here. And above all, do nothing without reading the manual first to learn about how your camera functions.

Gamma

Gamma is another variable for those wanting to alter the tonal balance of a camera. After you have read the Appendix to learn the background to gamma, you may not want to alter it at all. But film workers quite regularly use the equivalent processing facility to alter film tonal values. As long as you are able measure what has been done and restore it, there should be no problem.

Earlier in this chapter, mention was made of the relationship of gamma and 18% grey card; saying that, a broadcast specification camera reproduces 18% grey at 50% level. To do this requires that the camera has a gamma of 0.4. To increase the density of darker tones, raise gamma to 0.5. The actual value is unimportant but the effect is to set up the camera exposure to make the white card 100% as before and, using the camera manual, adjust the gamma to bring the grey card to 40%.

Obviously the gamma control must be accessible, and this is where menu control comes into its own, indicating how much you have altered the control. It may not be calibrated in gamma values but at least there will be a numerical value from which it has been moved and to which it ends up. This takes the worry out of changing settings. What value is decided upon must be determined by the picture and, like detail, can only be properly assessed in operational use.

Gamma values lower than 0.4 are not common, not so much because they have little pictorial merit, but because they can make the operation more difficult. Low gamma means high gain in the darker areas and this in turn makes the fill light in shadows quite critical. The matching of shots may then be compromised in shadow areas. At this point we must also take into consideration contrast control, the knee effect. This causes the higher tones to compress in rather a similar way to low gamma, although knee does not alter tones below about 85%. So the two effects can be rather similar.

Working with a high gamma, the grey card is at 40%, a useful feature for the more dramatic shots. Colour saturation will also increase in dark tones so beware that facial shadows do not become too florid, unless, of course, that is what you require.

Colorimetry

Colorimetry was an absolute 'no touch' parameter in the earlier days of television. It is not easy to measure the colour co-ordinates that are set into the camera and, as part of the colour processing, they are quite complex. Again, menu control of modern digital processing make adjustment less forbidding but it is still most important to know what you want to do before touching. Read the manual: gain some insight into what the camera does, for this is not a straightforward control to alter.

Generally speaking if a control is offered as a user-set facility it is in order to alter it. But always follow these simple rules:

Understand what the control does before you disturb it.
Make sure you are able to restore it to its exact original position.
Note – there are many controls inside cameras, particularly older ones, that are engineering set-ups and must remain so.

Menu control

In going through this procedure, we have checked the most significant parts of the camera. It may seem, to some, rather scant. There is a reason. What has been dealt with here is the basic camera system; modern cameras do not lend themselves to over-zealous tweaking. Just like their forbears, cameras need to be understood. And that starts with the basics.

One of the most innovative developments is the menu control through software of all the camera's parameters. The designer thoughtfully provided a reset action to put everything back to factory settings. There is little point in having the facility unless one is able to properly assess each change as it is made – which means understanding the picture. It's all down to experience; there's no short cut. Unless, that is, you are always willing to return to the camera designer's own ideas about your pictures.

When considering alternative methods of video measurement in a later chapter, the Vical method is described that uses visual techniques to compare tonal and colour balance against a known reference. This picture-oriented technique was created to fill the gap between the graphical waveform and the pure picture and provides simple measurement without its user ever becoming detached from the picture.

But it is a fact that all measurement methods have their limitations. WFM's, reference picture monitors and specialist devices; each has its own virtues and its own shortcomings. Neither will, on its own, completely satisfy all the needs of comprehensive video measurement. Photography, because that is what exists as the bottom line, is a combination of artistic and craft skills that we all learn in our own way.

CHAPTER 7

THE STUDIO

The previous chapter set out basic camera measurement. Now we shall look at a studio with multi-camera operation, plus, other sources such as graphics and caption generators. These are typical studio facilities requiring level set-up and timing.

Signal levels

Picture levels may exceed the standard limits. As described in the last chapter, a camera may be able to handle an extended scene contrast range that reproduces as an extended video signal. Levels as high as 125% are not unknown. Where shooting conditions are not fully controllable, out of doors with bright clouds in blue skies, extended contrast is an advantage. Modern CCD image sensors have a wide contrast acceptance and the signal processing is designed accordingly.

Where such programme material remains local and unlikely to pass through international circuits, such high levels may be acceptable. But it must always be remembered that these signals are not legitimate and if passed through equipment that removes any excess over the one volt, all fine highlight detail will be clipped back to 100%.

Peak white clipping normally takes place at the main studio output as part of the compliance with the video standard. It is worth checking out cameras to see if they exceed peak white, a procedure described in the previous chapter.

If pictures are allowed to exceed 100%, there's not a great deal of point in using colour bars as a signal level reference. It is confusing to those receiving your programme if your colour bars are not representative of the picture levels that follow. The recipient would be perfectly entitled to accept the colour bars as correct and allow the removal of any excess over the one volt. This applies in real-time programmes and recorded material.

It is for you, if you are the programme originator, to control your pictures and you must be prepared to work within the video standard.

Checking colour bars, as previously described, is part of programme origination.

Timing

In checking out the camera we did not mention 'time' once. Timing is a relative parameter and, for an isolated picture source, is irrelevant. But a studio with more than one source, has a timing requirement. The pictures must arrive in synchronism at the mixing or switching point if picture disturbances are to be avoided when selecting one source after another. Sources may be routed through diverging signal paths, through various video effects but must all arrive together at the same point in time. Those with longer paths must start earlier.

If pictures are to be mixed and inlaid together, they must be synchronised very accurately to ensure the picture elements register. When synchronism is achieved, a whole variety of picture effects and manipulation is possible; the output, however, always maintaining an unbroken sync pulse sequence.

Getting all the sources to arrive in sync at the mixer requires **genlock**. Each source must have the same picture starting point and to do so, its scan timing generator must be locked to all other generators. The simplest method is for one source to be master and slave the others to its output, a video input, called genlock, is provided for the purpose. Whilst this

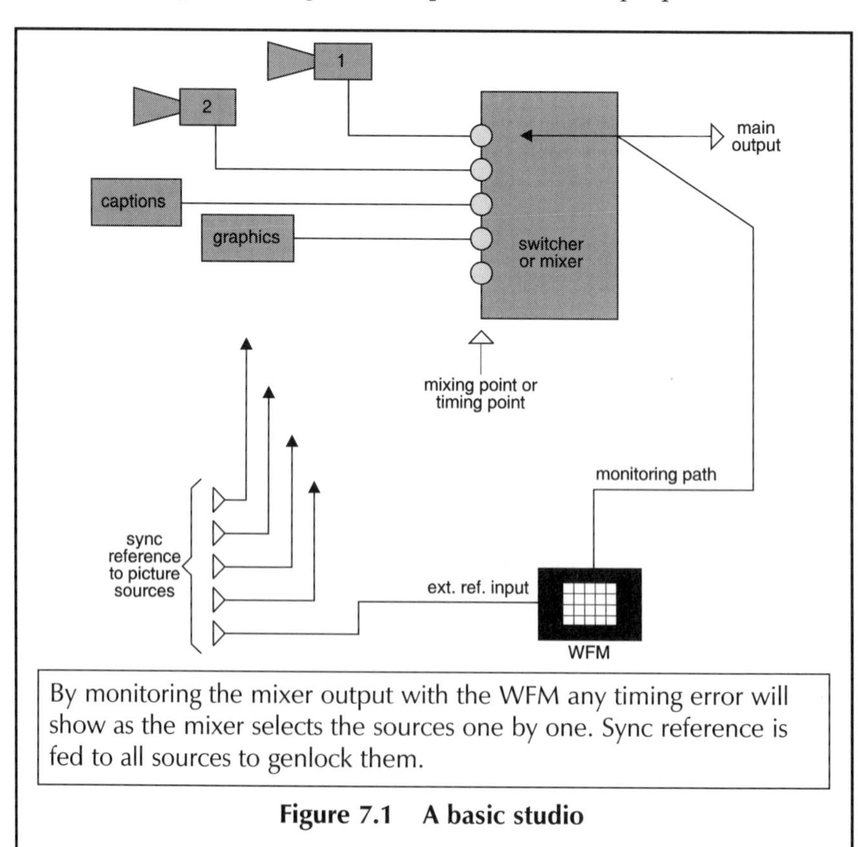

By monitoring the mixer output with the WFM any timing error will show as the mixer selects the sources one by one. Sync reference is fed to all sources to genlock them.

Figure 7.1 A basic studio

is perfectly acceptable technically, it is not always convenient to use a picture source and a dedicated source of genlock from a **sync pulse generator** may be used instead. The SPG, as it is known, is a reliable and permanent sync source of black and burst, often with additional facilities, such as test signal generation. Many installations make use of a test signal generator for this purpose as the freedom to change sources around is not compromised by one of them being used as the genlock source.

The simplified studio in Figure 7.1 shows how the mixing point is the timing reference for all studio input sources, the actual measurements being made at the mixer output. To allow alteration and addition to the system, a **jackfield intercept panel** can be introduced. This offers quick and convenient source interchanging but in so doing, the jackfield becomes a timed point in itself, with all video cables running from it to the mixer inputs required to be identical lengths. The increase in cost and complexity must be weighed against the benefits of increased operational flexibility.

By itself, genlock does not give synchronism. Genlock only makes the clocks in the picture source timebases run at the same rate. To ensure synchronous arrival at the mixing point requires that these clocks have to start at different points in time. The alternative would be to cut all the connecting cables to the same length, an impractical solution.

Timing adjustment is therefore provided at each source and may not prove very convenient. With the WFM placed at the mixer, its operator must instruct another to carry out the adjustment at each source in turn. Two people are required and if the operation is regularly taken down and rebuilt, as is the case with mobile productions, it can be a time-consuming business. Simpler fixed installations do work this way quite successfully, relying on the stability of modern electronics to reduce the amount of adjustment required.

Once a properly designed and installed studio is set up, the need for regular adjustment is minimised.

The control units shown in Figure 7.2 take different forms with various names. These form the technical apparatus that carry out the video functions. The camera control unit, or CCU, is sometimes called a base station; other units may be known as 'main frames', or simply 'crates'. By placing all the apparatus together, including that of the video mixer, a much more satisfactory engineering system is created. Measurement and adjustment now becomes confined to one area, often within the remit of just one individual.

Associated remote operational panels are smaller and lighter and more easily accommodated into confined control suites. Control of the apparatus remotely in this way simplifies design immensely and, quite often, the only video cables to enter these areas are those feeding the picture monitors.

In describing all this, we have made no mention of colour and it is now opportune to distinguish between component and composite studio design. With component working, the timing requirement is similar to

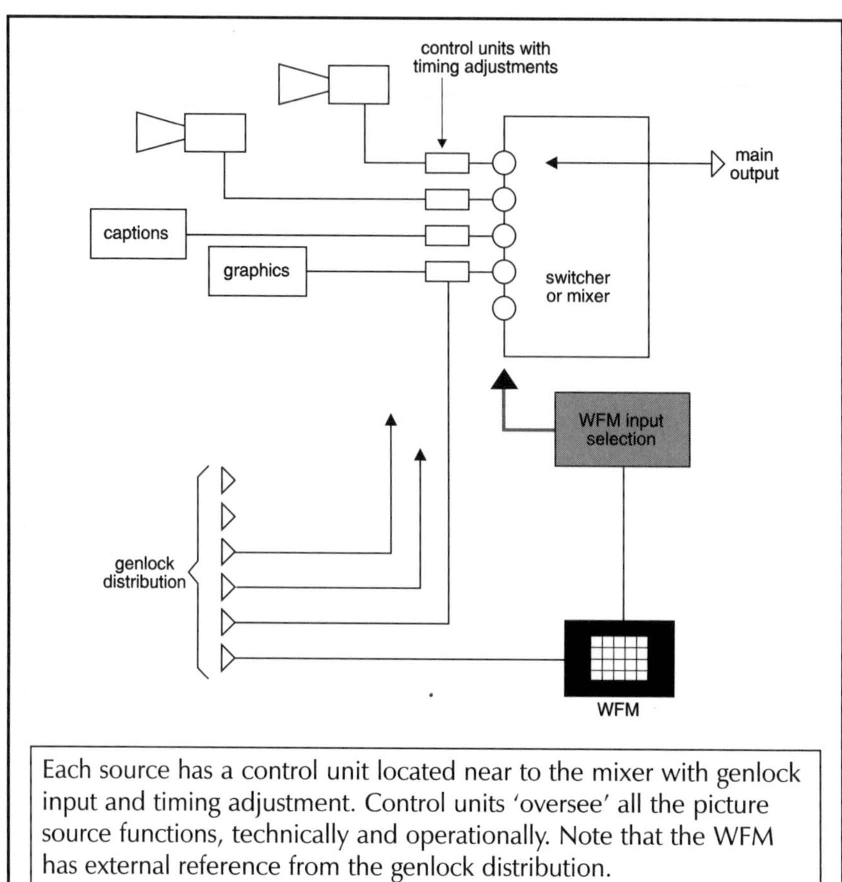

Each source has a control unit located near to the mixer with genlock input and timing adjustment. Control units 'oversee' all the picture source functions, technically and operationally. Note that the WFM has external reference from the genlock distribution.

Figure 7.2 A more comprehensive studio

that already described and conforms to what is called **monochrome timing**. The term 'monochrome' correctly describes 'one-colour', which is not black and white, or pictures without colour. Monochrome is strictly any colour, but only one, and therefore still constitutes a colour signal. But strictly correct terms become abused and where monochrome timing is stated it implies black and white or luminance timing. Timing requirements for component video are similar to that for black and white television, that is the synchronisation of the commencement of pictures, based on an accuracy of 0.1 μs.

Component demands accurate matching of the three circuits. Colour registration, that is the overlaying of the colour difference signals, must be done accurately if colour fringing is to be avoided. Registration of the colour and luminance signals is less demanding. If care is taken over design and installation, these requirements should not cause difficulties and once in place, measurement on any regular basis should not be necessary.

Composite colour timing

By far the most stringent timing requirement is for composite video. We have seen how the whole colour spectrum is contained in one rotation of the subcarrier vector. A metre of cable causes about 5 ns of delay or about 8° of vector phase shift. Bear in mind these are rationalised values for the subcarrier frequencies of both TV standards. Conduct a simple test in your own system, measure the phase delay of a length of cable by using a vectorscope. Figure 7.3 shows what to do.

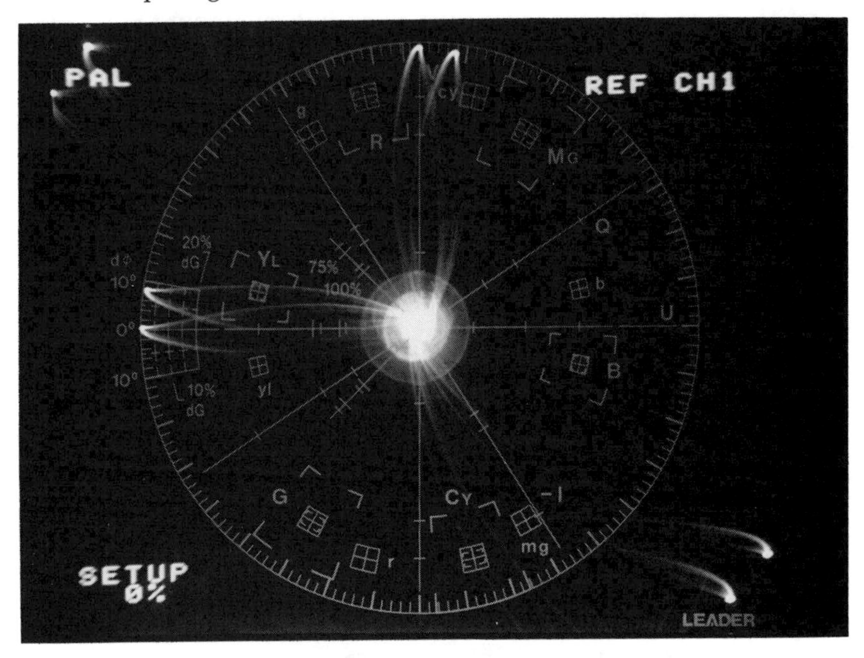

Feed the signal to input 1 of the vectorscope, then loop it to input 2 with one meter of cable and terminate. If the signal is PAL, the two inputs displayed together show input 2 lagging input 1 by 8°. For NTSC the difference would be about 6.5°. Any signal will do for it's the magnified colour burst that we are looking at.

Figure 7.3 The delay in one meter of cable

Referring back to Figure 7.2, let us see how colour phase affects this situation. In practical terms cable characteristics vary with temperature, age and wear and tear. These effects on a 50 m length would be significant over, say, a year's operation and the system design must take appropriate measures to deal with this.

A vectorscope is essential. It should be placed with the WFM using a common video cable to both units looping from one to the other. The combination forms the complete video test and measurement set, or just 'test set'. The length of cable running to these units is not significant, make it what is most convenient. But, always avoid damaged cable – this should always apply, and avoid altering the test set when once operational.

The sync reference may be black and burst or whatever is convenient as long as it is stable, reliable and representative of timing at the mixing point. It may be simpler to connect the test set to the mixer output; technically this is ideal for we measure the actual signals as transmitted. We will, however, be restricted to checks when the mixer is not in use, although there is no restriction on looking at the mixer output at any time. The mixer will invariably have an additional output to make this possible.

The dedicated multiple input test set is the most flexible but monitoring cables to the mixer inputs must be cut to identical lengths and must use the same cable type. Monitoring of the programme output is essential anyway for the confidence it gives in your operation, so, if a test set source selector is installed, make one of its inputs available for the mixer output. Timing at the mixer output is still the most true; checking here gives total confidence of timings and levels in the programme chain.

Figure 7.4 is a simplified system but illustrates comprehensive monitoring. The jackfield is a series of 'U' links that offer signal intercept points. Any additional source will be monitored and measured in the same manner as all the others.

Mixer timing procedure

The video mixer is a comprehensive control for picture cutting, mixing, fading and compositing (the inlaying of picture elements) used in the assembly of a complete programme for sending to a destination.

Where pictures are keyed or inlaid, mixed and faded to create a composite picture, more than one video source contributes elements of the signal. All these must be in perfect synchronism for luminance and chrominance. **Luminance timing**, or line sync timing, is termed variously as 'H timing', or 'H phase', sometimes simply 'line timing'. **Chrominance timing** is termed 'C phase', 'colour phase,' 'ϕ phase', 'CSC phase', or 'SC phase'. These are the usual terms applied to the timing controls referred to in Figure 7.2.

Control of line timing and colour phase are independent of each other, a further indicator that even in composite video, luminance and chrominence still remain discrete signals.

Where only one source is selected on the mixer, the whole of the signal will be passed, including the sync and burst. Where a picture is made from more than one source, the mixer must decide which sync and burst to use. How it chooses will depend on its design. It could insert its own black and burst, or require a dedicated feed. Black is a production requirement, as in a fade-to-black; the mixer output will then be simply black and burst. A source of black and burst may be derived from the

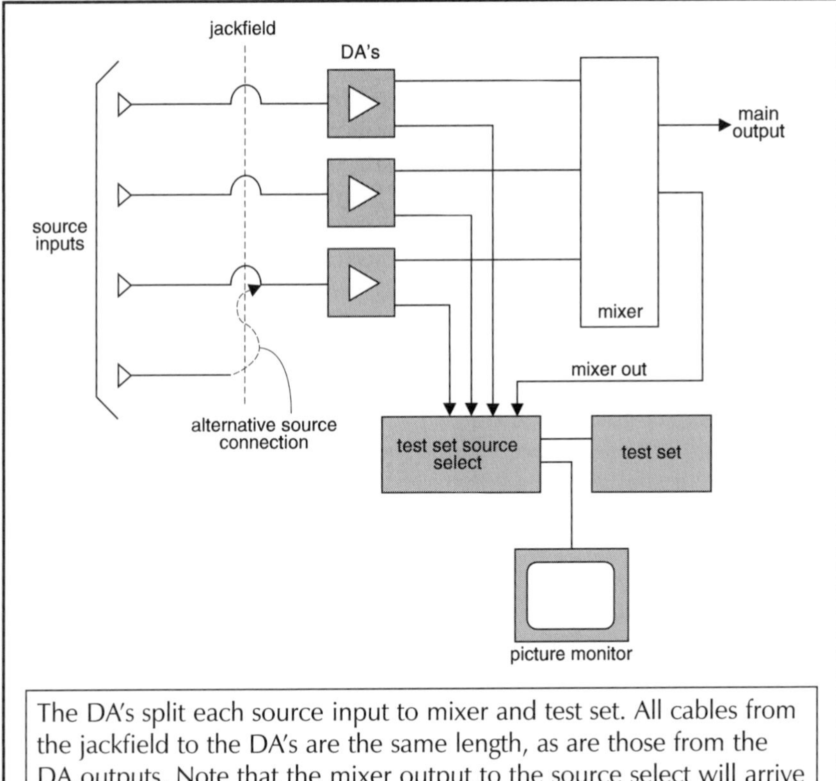

The DA's split each source input to mixer and test set. All cables from the jackfield to the DA's are the same length, as are those from the DA outputs. Note that the mixer output to the source select will arrive later than the sources due to the delay through the mixer and associated cables. Connecting an alternative source, as shown, will use an interlink cable longer than a 'U' link and will require that the new source is retimed.

Figure 7.4 Example of jackfield installation

genlock distribution. In some cases the mixer may itself produce black and burst for source genlock and for its own internal consumption.

Yet another picture source provided by the mixer, is 'infill'. Caption cut-outs may be infilled with a colour, or matte. Any other coloured mixer effect will also use internally generated mattes.

Now the mixer is becoming a signal source in its own right with its own black and matte generators. All are subject to the same requirements of synchronism as any other mixer input. These features are specific to individual designs; how they are dealt with will be evident from the manuals describing them. The basic principle of timing must remain intact, however, regardless of where the picture, or parts thereof, come from.

To carry out a timing check, start by choosing a reliable known source as a reference against which to compare. Use a camera and, for the moment, let us assume that it has a camera control unit. Control units detach the actual picture sources from the mixer; they act as interfaces and one advantage is to make the timing process simpler – we deal with the control unit, not the camera. The distinction is important.

With the test set, either monitor the source using the selector, or the mixer output with the chosen camera cut up, that is, selected on the mixer control panel. Many mixers have a multiplicity of paths for various effects but we want the main, or programme output. Usually designated PGM, all this does is to act as a simple source selector by depressing the button for that source and routing it directly to the output. It's well worth while having the mixer operational manual to hand.

Adjust WFM and vectorscope to place the displays correctly against the graticules, bearing in mind that its is H timing on the WFM and C phase on the vectorscope. When this is done, cut up all the other cameras in turn, watching their relative timings of line sync leading edges and phase changes of colour burst.

Note, there is no mention of colour bars on the cameras. It really does not matter whether pictures or bars are present or not as synchronisation only concerns line sync and colour burst. We are now dealing with the part of video which is outside the picture period. Timing is black and burst and nothing else – at least for the present.

Having said that, it's not easy for an engineer worth his or her salt to ignore colour bars, so switch them on and see how they look as you cut through them one by one. Differences will be due to individual camera encoder adjustments, circuit differences, or even individual camera design. Although not the point of interest at the moment, but it is good practice to make ourselves aware of how well the *whole* of the video signal performs.

These are the important points when checking source synchronism:

DO ...

1. Use mixer out if possible, otherwise, ensure the monitoring paths are correctly timed.
2. Use only line sync and colour burst for comparing synchronism.
3. Compare line sync leading edges at 50% amplitude; magnify to see clearly.
4. Compare phase of the colour bursts on the vectorscope; rotate and magnify if necessary.

Do NOT ...

1. Use unknown lengths of cable for connecting sources to test set.
2. Use picture information for timing measurement.
3. Alter any part of the test set during the procedure.
4. Alter the master source H and C controls during the procedure.

Should there be timing or phase differences adjust the respective H and

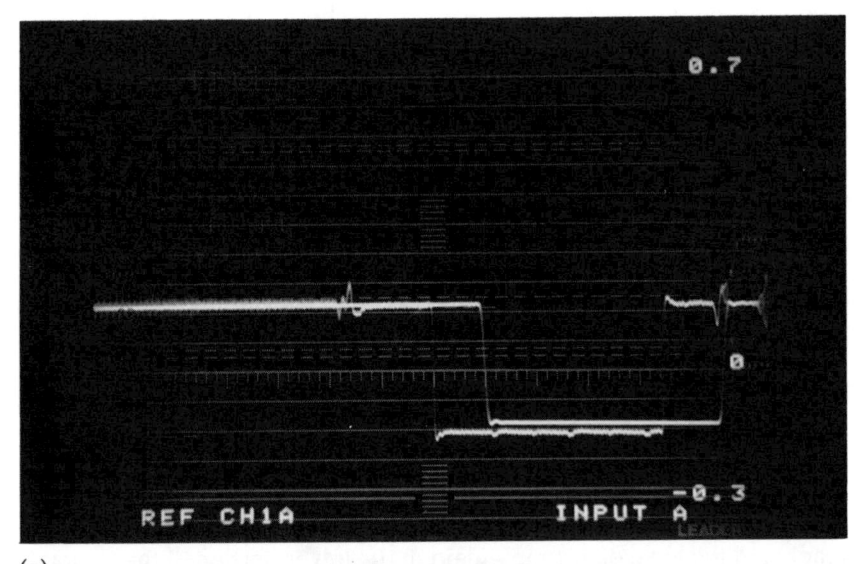

(a)

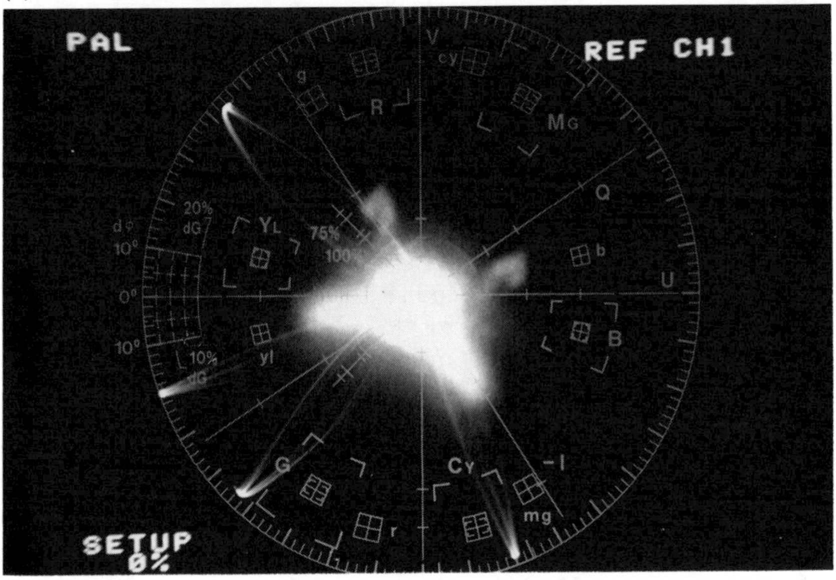

(b)

Two sources are shown together, one chosen as master. In (a) the WFM is magnified five times so that the line syncs are expanded. Timing is measured at 50% amplitude, by adjusting the trace vertically to bring this part over the timescale and horizontally to align against a convenient timing mark. The second source is 1 μs late as measured at the sync leading edges. The vectorscope (b) has the master colour burst aligned on its axis but magnified to reach the outer scale. The second burst is measured leading the master by 64°.

Figure 7.5 Timing and phase measurement

C controls on the control units, leaving the chosen master alone. If there is insufficient control range on any of the controls, it may be necessary to move those of the master and then go through the whole process again. This can arise where equipment differs in its design. As an example, C phase may have two controls, a quadrature switch that jumps 90° at a time, and a fine trim control. It is possible to get positions that fall between two clicks of the switch; a slight tweak of the master to move one way will give a more suitable position for the fine control. A good principle is to try and get all controls to end up away from their end points, or backstops. Once set up these should not alter greatly but a little adjustment range in hand is always nice to have.

Now for other sources. Graphics and captions can have the same treatment. The adjustments may differ, they may be via software, but there is no reason why this equipment should not respond in the same way as the cameras. The manuals will reveal the procedures to use in each case.

The situation shown in Figure 7.6 is of a composite image made from two sources, a camera and a graphic. The camera has inserted into it the graphic in the form of a 'strap' inlaid in the lower portion of the camera's picture.

Figure 7.6 is a simplified system but serves to illustrate to point about black and burst. Although some mixers may pass individual source syncs in simple cut mode, when compositing, this is not the case. Because the sources are switched in picture time it would be unwise to change over black and burst. If the graphic is 'see through' i.e. partially transparent which constitutes a mix, black and burst must be chosen from one or other sources.

Alternatively, the mixer may reinsert black and burst. Whichever way is used, H and C timing must, of course, be correct. It is quite possible for a phase (and sometimes an H error) to go unnoticed on the test set-up. The correct way is to time at the mixer inputs as in Figure 7. 2 or Figure 7.5. Here, the source black and burst H and C is checked, which will be representative of the *source picture* H and C. Checking at mixer out means that the black and burst of one of the sources has been discarded and, with it, its timing reference.

The picture is as likely to reveal how good the timing is, as is the test set. The eye will easily spot colour and saturation differences when checking between sources and composite. Timing can at this stage, be carried out using the picture but check what the test set is telling you whilst doing so.

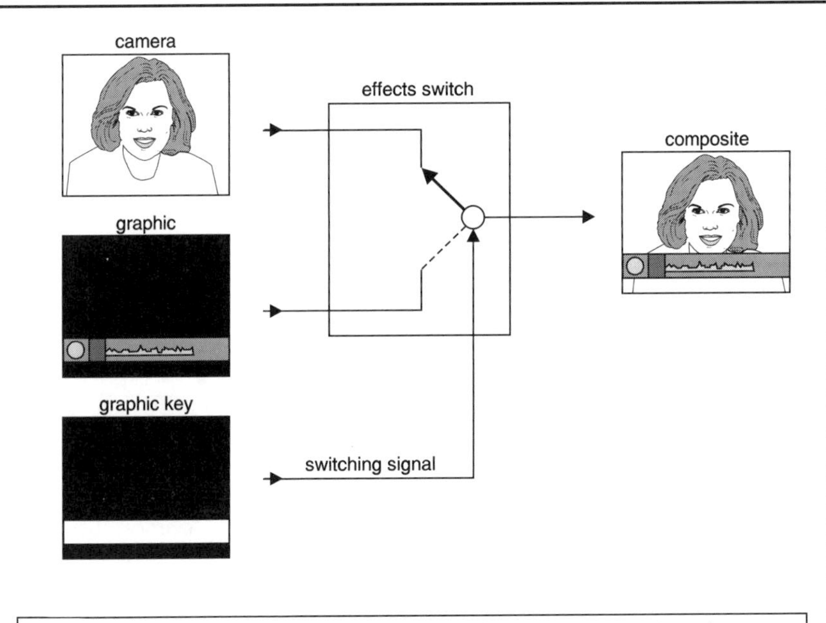

The camera has inserted into it a graphic. The composite has the camera's black and burst, and colour phase errors may not be apparent on the test set. Only by checking the sources then the composite will differences be apparent. On the picture monitor any phase errors will be evident from the difference in colour of the graphic in NTSC, or its saturation in PAL.

Figure 7.6 Phase errors in a composite

A remote camera

If there is a remote camera without control unit then it will require its own feed of genlock, presenting a potential difficulty. To get its signal back in sync with local cameras means that it must be advanced in time. Figure 7.7 describes the process.

Figure 7.7 describes the principle behind advanced genlock and how a camera can start its scan by referring to an earlier part of the waveform. Colour phase is dealt with in the same way.

No mention has been made so far about frame and field timing. That is because it is only necessary to identify line number; everything will fall into place from that. The genlock system compares waveforms at the field sync and identifies which field is which.

But there is more to the video system than just this as will be seen when we consider the mechanics of video tape recording. In the meantime, let's look at pictures.

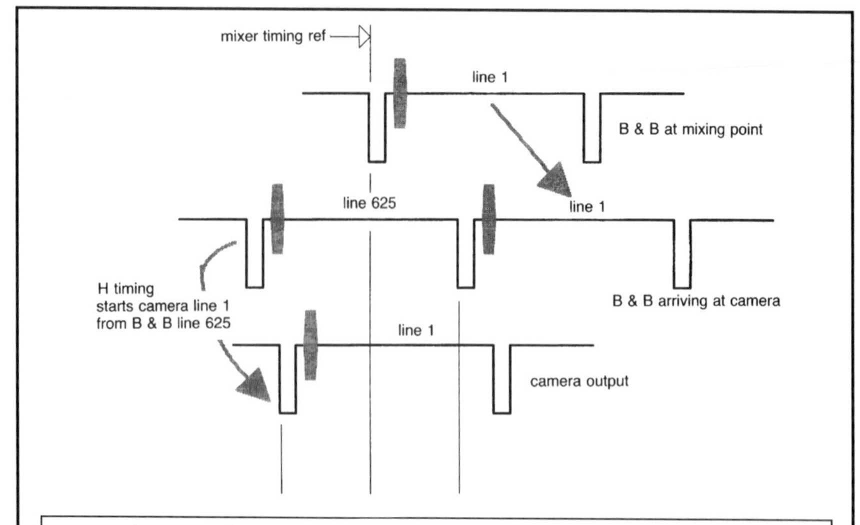

To synchronise the remote camera picture with other sources, it must start its scan early. Looking at Line 1, the black and burst genlock signal is delayed from t1 to t2 over the distance to the camera. Adjusting camera H timing to compensate demands that camera Line 1 is *started earlier* than Line 1 of genlock. To do this the camera generator refers to Line 625 of the previous field, starting camera Line 1 at t3.

Figure 7.7 Remote camera timing

The greyscale

Cameras are the most esoteric of picture generators, as we have seen already in the previous chapter. We need to have confidence that our cameras produce what we think they should. The greyscale line-up goes back to the start of television and, over the years, has been adapted to cater for one camera development after another.

It is not the intention here to set out the procedure in a step-by-step way. It is the intention to acquaint the reader with a knowledge and understanding that is adaptable to any situation and individual approach.

Nor is there a great deal of measurement, either. This is the business of *seeing the camera pictures* and using the test set as a confirmation of what is seen.

In Chapter 7 we used grey cards because they are an international standard. Greyscales unfortunately vary, but in a multi-camera system, where camera matching is more important than absolutes, a greyscale of any type is useful. It will show differences of colour camera to camera, at any level from black to white. Only one point regarding the greyscale needs careful consideration; its colour. What an unfortunate situation where greyscales are not grey, but this does happen. But there is a simple check and it's the Kodak Gray Card: compare this to the colour of your greyscale. You may even find disagreement about the colour of grey card, but that's really splitting hairs. Having looked at the greyscale, compared it to a grey card, and provided it is good enough to proceed with, now we must light it.

The operation of greyscaling is just like white balancing; the camera is told to make its output no-colour. If the input is not true grey then the answer arrived at will be incorrect and scenic and, in particular, skin tones will not be true. Studios operate with tungsten light, fluorescent, or other form of gas discharge lighting. Of these, tungsten is the most predictable as regards **colour temperature**. The Appendix describes colour temperature. Whatever the kind of light used in the studio, the lamp chosen to illuminate the greyscale must be representative of that used on actual sets.

The normal operating colour temperature of tungsten light is around 3000°K. Lamp brightness and colour temperature conflict with each other; changing the light output alters the colour temperature. The correct way is to set the colour by varying the level and then adjust the lamp to greyscale distance; or spot or flood the lamp, to get the required light level at the greyscale.

Light level

Light level too, must be typical of that used on set. Camera sensitivity now becomes important, the more sensitive the cameras, the less light level they require. But whatever the lighting system used, the greyscale illumination must be representative.

The light level chosen is as much influenced by air conditioning costs as lighting power costs. Light falling on the studio floor is a small fraction of the total power input, all the rest is heat. And that takes air conditioning power to get rid of it. Light level has other implications, for instance, depth of field.

The depth of field of a camera lens is dependent on its focal length. Also, because the sensor format is small the depth of field is large. Reducing depth of field is often desirable for it provides differential focus and the out-of-focus effect that are, for some productions, pictorially desirable. A lens iris setting of about f2.8 is a good aperture to aim for to achieve this. To reduce depth of field with modern sensitive cameras means low lighting levels, which is attractive because it saves money. But a sensible light level must be maintained for safety reasons. Cameras are also provided with neutral density filters to reduce light input and allow wider lens stops. Perfectly practical for the bright outdoors but wasteful of lighting power in studio.

Proprietary light meters are available to measure both colour temperature and light level. They are invaluable. Light the greyscale uniformly, place it at right angles to the cameras. Group the cameras so that their included angle is as small as possible. Frame each camera so that the greyscale is the same size in each picture. Measure the colour temperature, adjust the lamp fader to bring it to about 3000°K, or whatever value required. If possible avoid higher settings in the interests of lamp life. Light level should be similar to that on set and adjust by lamp position, or by spotting or flooding.

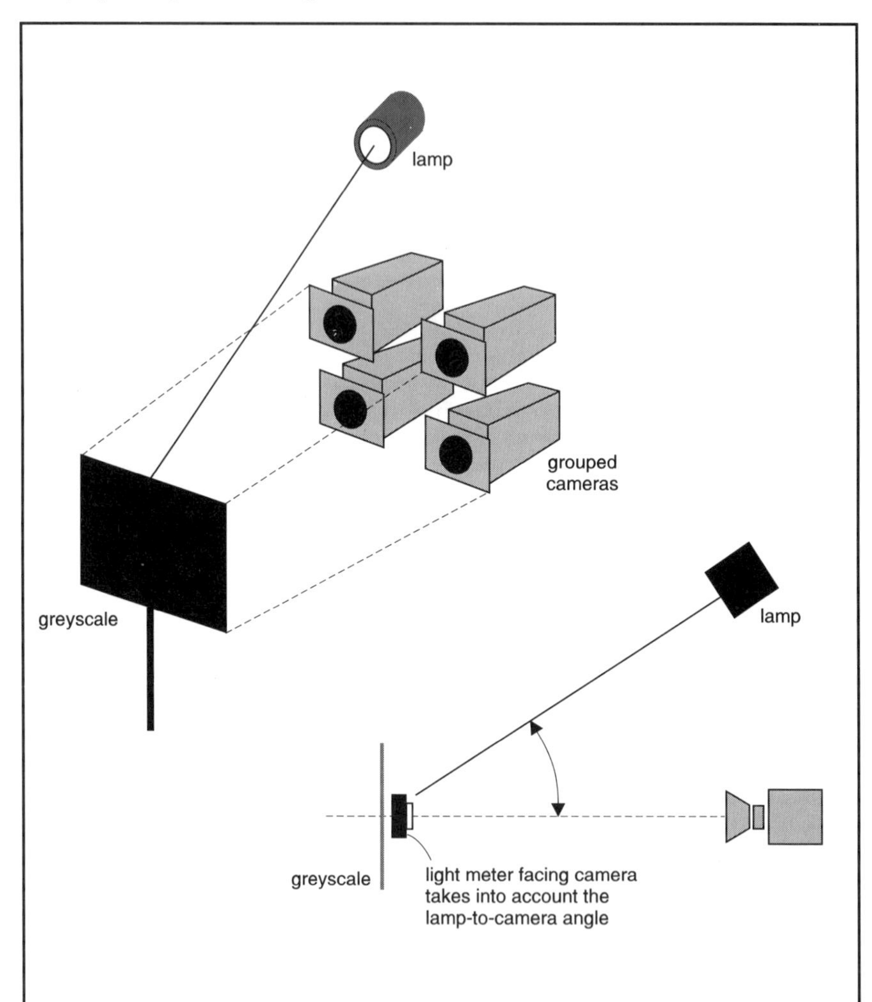

The cameras are tightly grouped and the greyscale is at right angles to the cameras. The lamp is at an angle such that any reflection from the surface of the greyscale is not seen by any camera. The greyscale is evenly lit. When measuring the light level at the greyscale, always face the meter to the cameras, not the lamp; that way you measure what the cameras see.

Figure 7.8 Cameras pointing at a greyscale

Camera control

In a multicamera studio, camera control is carried out from the control room. The camera operator has the usual control of camera position, lens angle and focus, but pedestal, iris, and colour are controlled remotely from the main production area. Sometimes a dedicated room is set aside for vision control and lighting.

The remote camera control panel will have:

Lens iris to control exposure.
Pedestal to set the picture blacks

These are mechanically part of the same control, a quadrant throttle-like lever is the iris. The rotary head of the lever is the pedestal, often referred to as 'black' or 'lift' in studio.

Colour gain; control of red and blue camera gains.
Colour black; red and blue pedestal.

These are the operational control panels, or OCP's.

There may be gamma, overall gain to set the sensitivity, dynamic contrast control, and other facilities as well. But the main ones, as set out above, are the basic controls to set up and match all the cameras accurately.

Firstly cap the cameras, that is iris all the lenses to close. Now pedestals can be checked – whether or not the nominal 3% is used is by the way – the object is to make all cameras the same. Use the picture monitor and the waveform and pay particular attention to relating the two. The picture will, at present, be black, but will reveal more about the cameras as each is selected in turn. Look for any slight colour variation in black, slight differences in camera noise, maybe shading errors as well. By operating pedestal and colour blacks these camera differences can be minimised. The object is to aim for perfection but with emphasis on getting them all the same, perfection may have to be compromised a little.

Choose one camera as master and adjust colour blacks until true black is achieved. Use the WFM, and vectorscope if available, and work to minimise subcarrier. The monitor obviously needs its set-up properly before the cameras can be looked at, but briefly switching the monitor to monochrome will show any of the monitor's own greyscale errors. Monitor set-up is dealt with in the Appendix.

Fine tuning

Having satisfied ourselves that the cameras all produce similar blacks, open the lenses. Expose to bring the white patches to 100%. Note what levels the whites of both wedges of the greyscale reach; it is almost certain that they will not be the same. Slight variations of light level will cause a few percent error, but this is not a serious problem as long as the same white patch is used for measurement of all the cameras.

Colour balance can be started by using the camera's own white balance system, then trimming differences manually with the colour gain controls. Should they fail to match at all from a white balance, there could be two problems. The greyscale may have insufficient light for the cameras to white balance correctly and some will make a better job of it than others. It may be that their internal set-up is mis-aligned.

Some cameras need almost white. It may not be good enough simply to over-expose to bring the grey background to 70 or 80%, for the over-exposed whites may cause the auto balance to abort. If the result is still failure to match, at least reasonably, then set-up errors are almost certainly the cause. Now there is a dilemma. If this part of the camera operation is at fault, it begs the question of what else is wrong. One can only advise diligence from now on for other conditions that may appear.

Manual adjustment will usually be successful; adjust colour gains until true grey is achieved, not in white, but lower down at about 60 to 70%. Check all the levels of the greyscale for colour, used picture and test set.

Check the peak white clip points, over exposing by 10% or so. Check for colour differences in the over-exposed whites and that the clip level is 100%. All cameras should be the same here. Use both picture and waveform to assess.

Now vary the irises from closed to fully exposed and watch vectorscope for the display shifting out from the centre. Relate what you see to the picture. Push into over exposure and observe again. Note how the cameras differ in their response. This is a very thorough test.

Engineering

If there are set-up problems that cause differences camera to camera they will be a perpetual nuisance in the operation.The OCP controls specific sections of the camera, if set-up errors exist elsewhere, any compensation made by the panel will be limited and the errors may emerge elsewhere. They really need to be put right so, yet again, we resort to the camera manual; this time it will be the workshop manual. And that's engineering!

Engineering is not a rite. It is a perfectly proper means to an end, that of having the equipment functioning properly. Do not be afraid to investigate faults. It may be no more than observation, measurement and a carefully written fault report.

Taking the side off a piece of equipment – any equipment – is a big step and some knowledge is necessary before you do so. There is also a question of safety, not just yours but that of your colleagues as well; that is your *personal liability*. This is where **training** is so valuable: to know what you are doing. Training is fundamental to television and video just like any other discipline. If you don't know, don't interfere.

Correctly working facilities mean reliability and confidence. They also mean an easier life for operators and programme makers alike. That's all part of good and efficient working practices.

CHAPTER 8

VIDEOTAPE

The studio described in the last chapter employed video sources that were predictable in their timing. Videotape players are not, they are mechanical systems that have irregular timing variations, or timebase errors.

All tape machines suffer the effect to some degree and various ways have been developed to correct the condition. All involve electronic correction.

In Figure 8.1 the VTR has a genlock input but this need not be complete black and burst as the machine cannot maintain line sync accuracy, and colour phase will be completely disregarded. The machine will maintain lock in frame only, that is 2V; hence this is called simply 'framing'.

Paradoxically, the advancement in electronics has permitted further degradation, or tolerance of timing accuracy, of the mechanics. Whereas

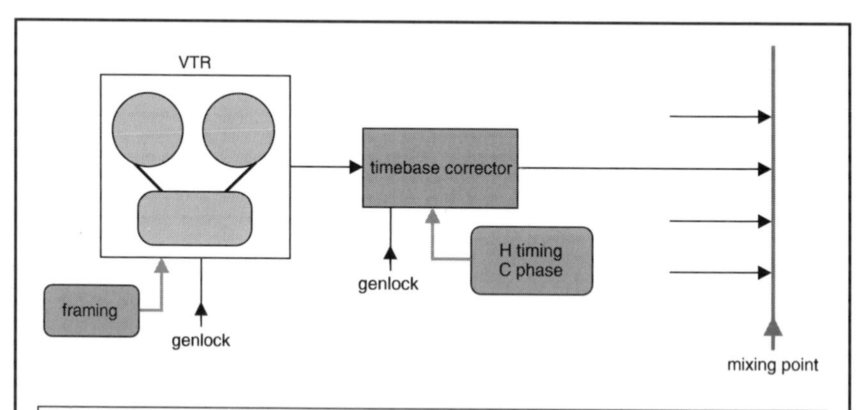

The timebase corrector (TBC) holds enough signal off the tape in its memory to cope with timing variations of the VTR. If the signal jitters by 1 ms, the TBC will have to hold at least 16 lines of video. Genlock is applied to the VTR but this only maintains an approximate timing centred on the mean jitter variation. The TBC genlock accesses the memory allowing precise timed release of the signal. Pictures off the VTR will start about 16 lines early and will be subject to a TBC delay of the same amount.

Figure 8.1 VTR and timebase corrector

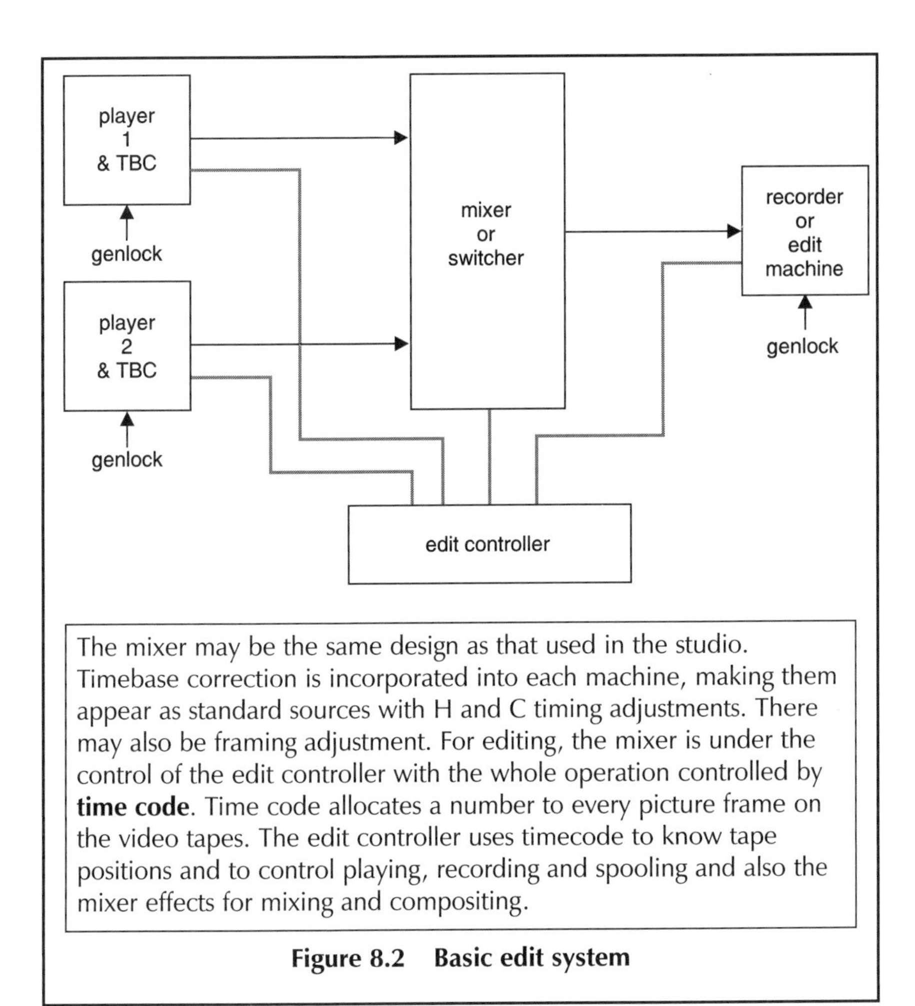

The mixer may be the same design as that used in the studio. Timebase correction is incorporated into each machine, making them appear as standard sources with H and C timing adjustments. There may also be framing adjustment. For editing, the mixer is under the control of the edit controller with the whole operation controlled by **time code**. Time code allocates a number to every picture frame on the video tapes. The edit controller uses timecode to know tape positions and to control playing, recording and spooling and also the mixer effects for mixing and compositing.

Figure 8.2 Basic edit system

a timebase range of tens of lines was a norm, we now talk in terms of frames. From a line memory store to a frame memory store and so the synchroniser is born.

The synchroniser

To digress from the tape scene for a moment, it is worth briefly looking at the synchroniser. It is a universal timebase corrector, because it holds a complete picture in its memory. It may be put to a number of uses from VTR synchronisation to video effects.

Put simply, it is a box with a video input and video output. It has enough memory to hold a complete frame. A second input is for the reference, usually black and burst. This controls the memory output to bring the source video into synchronism. A synchroniser can make up for clock rate differences, meaning that a distant source without any timing reference to the mixing point whatsoever, can be made synchronous. The output appears the same as the input for the device is transparent, apart from the timing change. But in so doing, the signal is delayed by one frame.

Such a delay is, in itself, of no importance; the machine just starts its picture output that much earlier. But one must consider the other part of the videotape output: audio. The audio is not subject to a timing delay and will therefore be advanced over the video. This may not be a great deal in audio terms, or lip sync, but where multiple synchronisers are used in a complex programme chain, the audio will require delaying by the same amount.

Synchronisers are used mainly for distant sources but the versatility they offer make their use on mixer inputs more common. Genlock now becomes local to the synchroniser and any source may be connected regardless how far away it may be, physically or in time.

The same timing controls apply, that is H and C, with the addition of signal level controls as well; gain, lift (the engineering term is preferred to pedestal) and saturation.

SC-H phase

Videotape brings in the final element of composite video. So far we have taken the chrominance and luminance as separate signals added together to make composite video, a model that is not strictly correct. Subcarrier is the master signal. It is generated by an accurate clock, quartz in most cases but in large scale broadcast systems, it is many times more accurate. Rubidium is one example.

Line frequency, 15.625 kHz in PAL and 15.734 kHz in NTSC, is derived precisely from the respective subcarriers. Field rates are determined by the number of lines and length of line; they are 50 Hz in PAL and 59.94 Hz in NTSC. By fixing all parts of the timebase to a single master in this way, the detail of the relationship is always correctly maintained. If the frequency of subcarrier changes, the line and field frequencies change accordingly, so retaining the integrity of the signal for it is most important to keep the relative timings and positions intact.

Chrominance and luminance are therefore related very closely indeed and cannot be separated and rejoined without due consideration of subcarrier phase and line timing. The relationship is known as 'SC-H phase'. The specification states the subcarrier phase-to-line timing relationship and is set out in Figure 8.3.

After 4 fields (2 frames) the subcarrier is back in phase with the start of field 1. NTSC is therefore a four-field sequence.

For studio applications SC-H phase is dealt with as part of the sync pulse generation system and all picture generators will be locked together maintaining the 4-field sequence. Video tape is less predictable

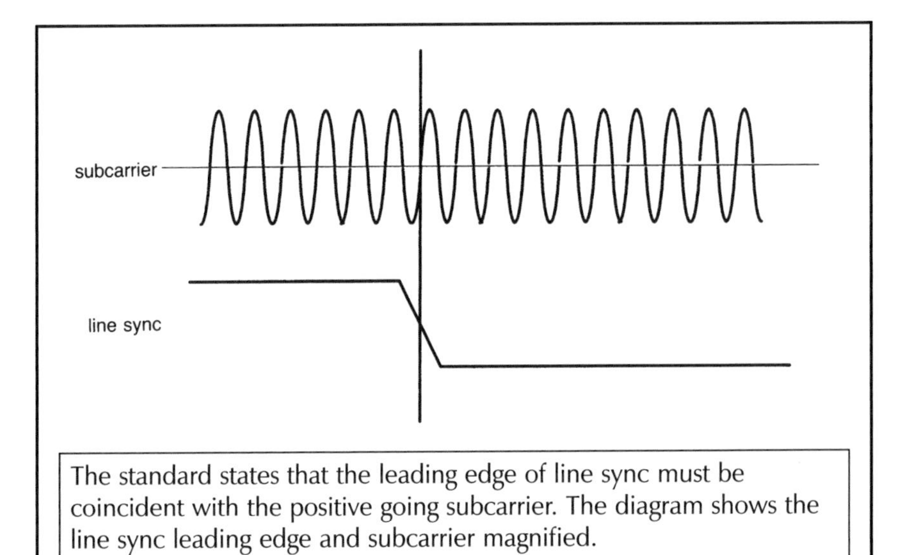

The standard states that the leading edge of line sync must be coincident with the positive going subcarrier. The diagram shows the line sync leading edge and subcarrier magnified.

Figure 8.3 SC-H phase

as tape speed is mechanically dependent. Each tape player must be aligned to its neighbour in the 4-field sequence.

Should field 1 from one player be aligned with field 3 from another and edited together, the fields will match but the subcarrier phase will jump by 1½ cycles over the edit. The odd half-cycle is a reversal of colour phase and the picture display compensates (a brief colour change may be seen) by shifting line timing to make good the SC-H error, resulting in a picture hop sideways.

More differences between PAL and NTSC arise with SC-H phase. The differences in field rates and the number of lines in the two systems influence choice of subcarrier frequency quite markedly. The reasons for this will not be gone into here, only those directly concerning the discussion. The implications of SC-H phase as set out in Figure 8.3 may appear at first sight to be insignificant; a cycle of subcarrier is only one-twentieth of the duration of line sync. We are interested in H timing and C phase so does it matter if these become shifted in relation to each other? Why should so small a timescale trouble us?

Let us see what SC-H phase means.

An alternative name is 'colour framing' which goes some way to describe the condition. The frequency of subcarrier was decided upon when black and white TV was well established and it was considered important not to disadvantage viewers in black and white by the addition of a colour signal. Subcarrier would be visible on black and white receivers as a dot structure, and how serious this was depended on the frequency chosen. Both standards came up with different frequencies. Field rate, number of lines, and, in the case of PAL, the alternating colour phase for every line, all influenced the final outcome.

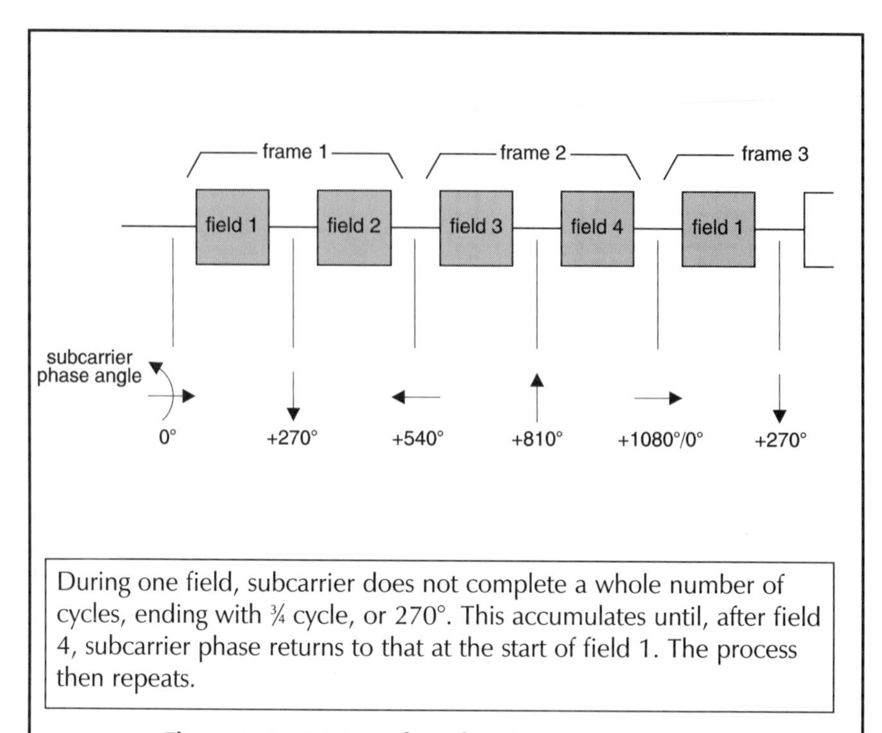

Figure 8.4 NTSC colour framing or SC-H phase

NTSC has the least complex SC-H relationship and so this will be considered first.

The number of cycles of subcarrier per field is:

$$\frac{3579545 \text{ cycles/s}}{59.94 \text{ fields/s}} \approx 5918.75$$

The '0.75' in the answer is three-quarters of a cycle, or 270°, which means that over one field, the subcarrier does not complete a whole number. The sequence may be set out as in Figure 8.4.

Colour framing in PAL is twice as complex. PAL subcarrier is 4.43361875 MHz. giving:

$$\frac{4433618.75 \text{ cycles/s}}{50.00 \text{ fields/s}} = 88\,672.375 \text{ cycles/field}$$

An odd number of quarter cycles is indicated by the '0.375' and results in the PAL colour framing having an 8-field sequence. That is four frames before subcarrier phase returns to its starting point, making colour framing in a PAL edit system that much more complex. It may appear operationally quite restrictive; four frames is a significant time span where cutting points are concerned. But once the VTR players are set up in the correct framing sequence, the actual cut, or edit, may be anywhere and the sync pattern will still be maintained correctly. It is, however, restrictive when one considers that the play in pictures can only be aligned on this 4-frame basis rather than a single-frame basis as is the case with component video or film.

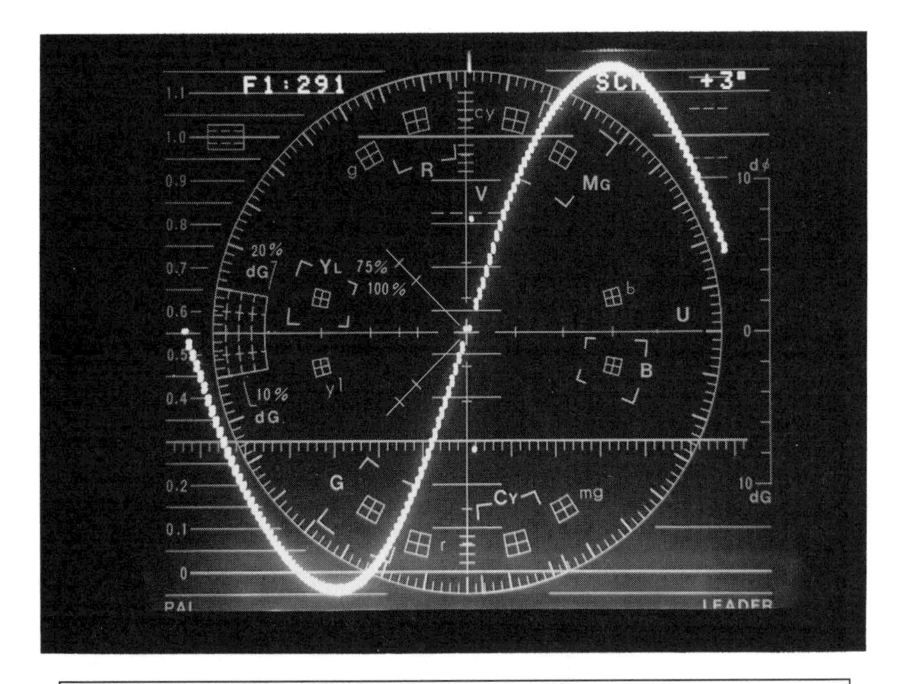

The complexity of the measurement has meant that various methods of display have been developed. This example places the crossing point positive-going subcarrier at the display centre. Here, the line sync leading edge, depicted as two dots above and below the crossing point, are very close to coincidence. The numerical display states the error is +3°.

Figure 8.5 Measuring SC-H phase

Measurement of the parameter is not simple by conventional means because the time scales of subcarrier and lines and fields are too great to display on a conventional scale. It is, therefore, necessary to have special facilities to make the check possible. Figure 8.5 shows one example.

Colour framing, in practice, may be thought unimportant; a hundred or so nanoseconds is only a small hop. Maybe we could dispense with the hindrance of working in a four-field sequence – in PAL, that's eight fields – and ignore the effect to gain a simpler installation and operational speed. Colour framing does add quite a lot just to get perfect, invisible edits. Or is there more?

If we ignore the SC-H phase requirements we run the danger of some equipment rejecting our signal, or worse, attempt a cover-up, causing confusion elsewhere that may not be discovered until the programme is replayed. The standard is written for a reason whether it is convenient to all users or not. Television transmission is too complex to allow one signal to compromise the whole system and there is continual assessment throughout the chain. The whole system is integrated; each part depends on all others doing their job correctly – and that includes the users of it.

SC-H phase is the final element of the synchronising pattern of video. It is the most complex with its origins way back in the development of colour television and the need, as perceived then, to retain compatibility with monochrome receivers.

Today, far more use is made of the component recording. Although this increases complexity with its tri-circuit demands, the advantages are the avoidance of complex synchronisation and much improved picture quality. Although the subcarrier system was specifically designed to resist colour distortion, there comes a point when the system fails. It is the continual replay and recording that takes place during editing that particularly affects the signal.

Generation loss

Playing back and re-recording causes a quality loss in the signal. The degradation occurs at recording and again at playback, an effect arising from a number of factors. The electrical signal is relatively free from fault during these processes; it is where the signal passes to and from the tape that errors arise. Magnetic distortions are complex and corrections only nominal. The signal, at this point is wrapped into a high frequency carrier envelope before recording, to afford some protection.

But, with all the current technology available, distortion still occurs. Each time a signal is recorded it is called a **generation**. In a typical editing operation two, three and four generations are quite usual. The condition is most noticeable as noise interference with subcarrier showing as noise graininess in well saturated colour areas. On the WFM the trace thickens and what was once a clean distinct line, this is now not the case. The vectorscope too shows the condition as chrominance which is seen to spread out and becomes difficult to measure accurately.

Another related condition is **dropout** and this affects composite and component systems. This is where the tape has physical flaws in its makeup that causes the signal to disappear. Dropout may be only tiny blemishes or extend over whole blocks of lines. So serious can be the effect that compensation has been used ever since VTR became established. This operates by inserting the previous line, or part of it, into the gap.

Component videotape's greater resistance to generation loss has meant its almost total acceptance in many areas of television. But where network transmission is concerned, composite is still the standard.

Setting up for editing

Let's start with the recording, or edit machine and make a test recording of colour bars. The source of these should be a local generator and checked to make sure they are good. Before recording make sure the machine is in the correct mode, check with the manual about the TBC and record settings. These will be subject to an engineering procedure that will optimise them. There may be specific control positions referred to in the manual and these should be adhered to. Record about half a minute, or more, of colour bars.

Play back and check the result. Adjust the playback controls until the bars replay are correct. The controls are:

1. **Gain.** Sets overall signal level or amplitude. There may be a separate control for sync amplitude.
2. **Lift.** This is pedestal or black level.
3. **Saturation.** The amplitude of chrominance.

It may be questioned as to why it is so important to set up the replay side of the record machine. The edit monitoring is from this machine, anyway, so how else can we check our recording?

Playback of composite video will look distorted; recorded colour bars won't look like the colour bars from the generator. Chrominance will not have the clean subcarrier envelope, but the white bar should be clean enough to measure. Set it to 100% with the gain control. Adjust lift so that the black bar is at 0%. Make sure it's not below, it can only be set properly by raising, then carefully setting back to black level. Do not, at least at this stage, put in any pedestal unless it is part of your standard operation.

Saturation is last: it must be set to bring the lowest edge of the green bar envelope to 0%. Obviously lift must be set before saturation. It is good practice to check all three again; it only takes a moment – particularly as the recording degradation will cause some difficulty in precise adjustment.

When everything is satisfactory, 'stripe' the edit tape. The edit tape takes the edited programme and black and burst is recorded on the whole length of an unused tape. In the process, timecode must also be recorded so that the edit controller can recognise the tape position and can therefore control the edit machine.

All recorded programmes should have colour bars on the first 30 seconds of tape which will be used to set up subsequent replay machines. It is always assumed when accepting recorded material that the bars' preceding programme are correct, hence the care attached to recording them in the first place.

On the play-in side, signal levels are not predictable and play-back line-up should be done whenever a tape is changed. This is imperative as recorded source material may be incorrect, tape performance varies and the recorder itself is an unknown. Recordings are outside the control of the edit system; they could be sourced from anywhere and on any equipment, leading to considerable variation in all parameters.

Also, where the source recording is a camera, bear in mind the signal levels of actual pictures. In the two previous chapters we looked at how cameras often produce signals exceeding 100%. The camera colour bars are a standard and will not indicate picture levels, so check a portion of the tape where there are highlights present, windows or bright skies. Check to ensure these do not exceed 100%. Small highlight specula that exceed 100% are less of a problem than larger areas. The mixer, or **proc amp** that follows it, should limit to 100%; adjustment of peak white clipping might be available but should always remain at 100%, or 0.7 V above black. Do not be tempted to adjust higher to pass the excesses of the camera.

Controlling high camera levels should be performed by reducing playback machine gain. More ideally, if a colour corrector is available, use whatever combination of control it offers that is appropriate to the picture. Colour correction is an extension of the photographic process and a very specialised operation. It should be done with care and in sympathy with the programme as a whole.

Carrying out a player line-up is the same as that just outlined for the edit machine replay. It deals with the player and source material together and, in fact, the player may become offset to accommodate source recording errors and tape variations. The degree of offset may not always be apparent unless you are familiar with the equipment. It is worthwhile studying carefully the off-tape colour bars to see if any other peculiarities exist. A later chapter looks at common distortions.

Playback H timing and C phase adjustments follows similar procedure as for studio source line-up, except that colour framing must be considered as well. Timing adjustment controls the TBC's, effectively making it independent of the tape signal. Once set they are unlikely to require further adjustment, unless the system is re-configured for any reason.

Because the machines are variable, a standard must be used against which timing comparison can be made. A dedicated mixer source is able to provide such a reference; sometimes this would be colour bars 'hard wired' as a mixer source with fixed H and C timing. Alternatively, a graphics or caption source may provide the reference, although these may become subject to operational changes and, therefore, less reliable as a timing reference. It is common practice to establish a timing procedure for a particular installation depending on the operational requirements.

As composite editing is replaced by component, so synchronisation becomes that much simpler. But composite operations offer considerable flexibility of sources and manipulation. In the process known as **conforming**, original component masters are edited to composite for transmission. The edit machine may be a composite recorder, with component play-in machine outputs encoded. There is now a colour framing requirement. The edit machine will sometimes indicate if colour framing is incorrect, but not necessarily say what the error is.

The timing point is the mixer as before. The many variations of edit system make specifying procedures difficult. One can only refer to operational manuals for detail information.

As a final check of machine timing we can utilise picture information: set up a horizontal wipe on the mixer, between two play in machines. Figure 8.6 explains further.

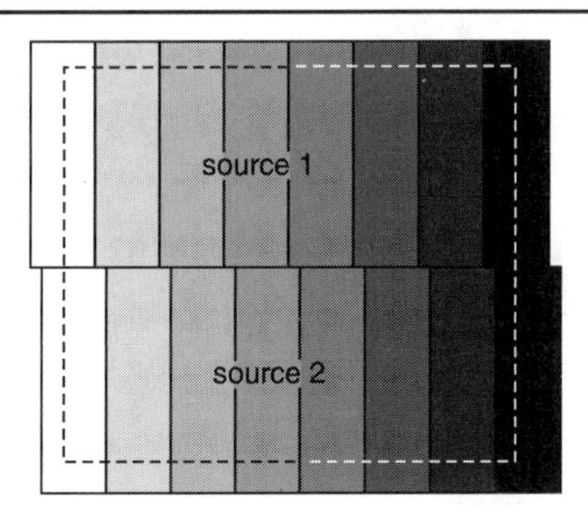

In this diagram of a picture monitor, the mixer horizontal wipe is shown in screen centre and the picture is of the mixer output. The dashed rectangle is the limit of the screen mask. Both machines play colour bars split by the mixer to appear top and bottom of the wipe. The wipe is an example of where the mixer decides which sync and burst to use, it cannot pass both. In the example shown here, the mixer uses sync and burst from source 1. The timing of source 2 is late and is consequently displaced to the right.

Figure 8.6 A horizontal wipe showing timing error

Unlike the studio situation described in the last chapter, where only black and burst was used, for this check, colour bars provide accurate picture timing when viewed together in a wipe configuration: an example of how a picture monitor can show timing differences although unable to make timing measurements. H timing of source 2 is adjusted to make the bars coincident and is picture timing, not syncs.

In practice using the picture in this way is common and an excellent way of getting coincidence of picture timing. There are sometimes discrepancies between line sync edges and start of picture, a consequence of variation in blanking duration. It is unwise, however, to place total reliance on the picture – monitor, vectorscope and WFM all have equal parts to play: use them together. When mixing and compositing, picture timing is important and the final trim is often done at the expense of line sync timing. Pictures are the final arbiter but when making these judge-

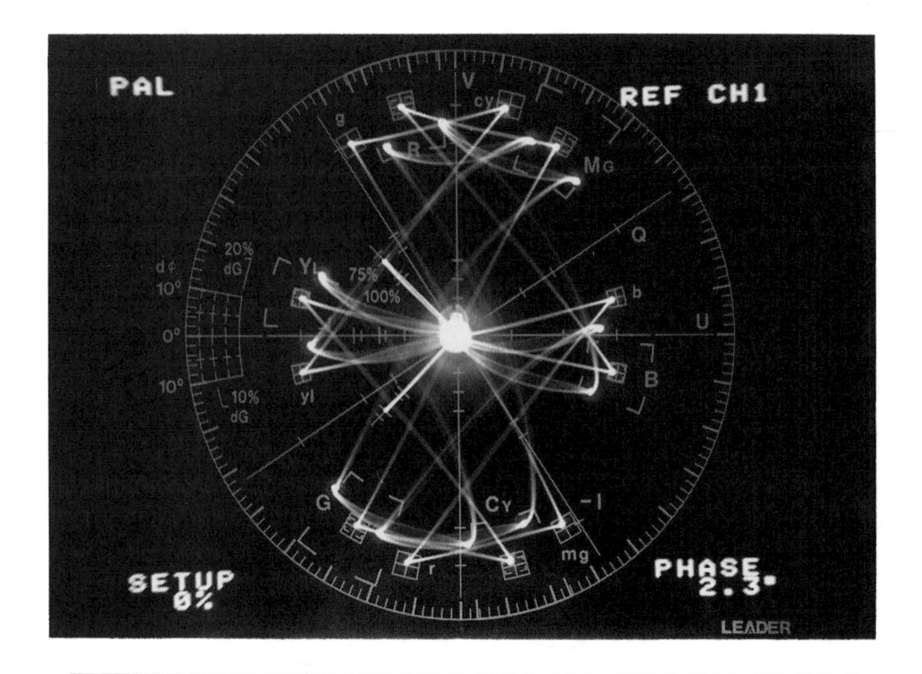

The situation is as in Figure 8.6. Only one burst appears, that of source 1. One set of colour bars is correctly phased, the other leads by 10°.

Figure 8.7 Horizontal wipe showing phase error

ments do not ignore the implications of deviating from standard timing requirements.

In the same way, phase error between the play-in machines will reveal itself as two set of colour bars on the vectorscope. Rotate the display to position the colour burst; note that one set of colour bars is in the boxes the other, outside. Adjust C phase of the offending machine to correct; you may not be able to identify which machine is in error until you adjust one of them. If both bars are wrong, the mixer is putting the burst where it thinks is correct and you therefore must assume that *both players* are mis-timed.

Now carry out the same procedure whilst looking at the colour monitor. As you vary C phase on one machine, you will see the colour bars from that source change in saturation in PAL and in colour in NTSC. The PAL system compensates for the phase error but desaturates. The monitor may have a facility called 'delay PAL'. This selects between simple and conventional PAL. Simple PAL is less sophisticated and does not desaturate but allows the alternate lines to show the phase error, one leading, the next, lagging. At a distance the eye integrates and effectively carries out the desaturation process itself. Looked at more closely and the alternate line phase error will show. When the C phase for both sources is the same, both halves of the screen will be the same.

When satisfied that all appears good, try a test edit between a play-in machine and a local source, such as captions or graphics. Play back the result: we are doing the ultimate check-out to see if the edit is accepted. If anything is wrong with our line-up it should appear now, there may be a glitch on the recording as the edit passes through, a flash of colour change, or a warning indicator. Repeat and watch the test whilst monitoring mixer, or edit machine outputs.

Videotape recording is a complex operation and requires the whole system to obey the standard. If it is set up incorrectly or a fault exists, recording will not be correct. It may appear when the playback is checked, or worse, it throws up a problem on another machine. If that machine's a hundred miles away, it's too late. Watch out for anything untoward as the edit proceeds. If the source tapes are swapped it will be necessary to go through all these checks again, although timing and phase should be taken care of by the machine TBC's.

The more the procedure is gone through, the quicker it will become; a forgotten or sloppy line-up is an embarrassing waste of editing and production time. Be methodical and develop an operational discipline. Experience is so valuable here. Knowing waveforms and the equipment, what it can do and what it will not.

Component working cuts a great deal of the complication of timing out. We have still to 'H time' and levels must still be checked, possibly including each of the individual component levels as well. In many cases the monitoring will be composite although the main signal paths are component. Level variations between the components will show as composite colour or saturation errors on colour bars. For this to be acceptable the composite encoding and decoding must be above question. There is only one way to guarantee that this is so, and that is to check it, or have it checked.

We may feel that component is a blessing, indeed it is where tape degradation and timing is concerned, but there are three signals that *must match perfectly*. Keep an eye on the picture for colour changes, they may be subtle, but could point to internal machine line-up. It is unlikely that the measurement test set will indicate the small errors that will be evident on picture.

CHAPTER 9

DISTORTIONS AND TEST SIGNALS

Distortion takes place whenever a signal passes through a circuit or piece of equipment that causes unwanted change to it. The action must not be confused with intentional alteration such as pictorial alterations.

We have already used the universal test signal, colour bars, extensively. The ramp is another useful test signal; it is easy to generate accurately and its form quickly shows linearity changes. Both these test signals have limitations, and some of these have already been discussed.

There is a great deal of specialised measurement that falls outside the remit of this book; it is a subject on its own for engineers involved in design and set-up installations. But as some test and measurement equipment offers a variety of test signals and, because it's useful to know something of the subject, we will look at a few of the features these offer.

Test signals and distortions naturally go hand in hand. In many instances a distortion will not be evident until a test signal identifies it. That is as it should be, if its severe enough to be seen on picture, it's usually too late. One would hope a check-out beforehand will identify a fault before it becomes a problem. Colour bars will uncover many faults if used properly. It is really a matter of careful observation, not just checking the white bar is at 100% and the green envelope reaches black level. Practise in observation will alert one to nonlinearity problems, poor frequency response and differential phase distortion, to name just a few.

Faults, unfortunately, can appear at the most inconvenient times, although there is often some warning. A quick check through the system before the day's work will be worth it if a pending problem is spotted. A good picture and an experienced eye, good test gear and thorough understanding of the system, are the bases of video operations.

Distortions arise from faults and incorrect operation. Faults arise because equipment wears out or is not used properly; it may be subject to excessive heat (or cold) or moisture, or maintenance may have been inadequate. Incorrect operation results in equipment working outside its design specification. It may not be compatible with the rest of the system, or it may have signal demands made of it that are beyond its capabilities. The list can go on.

Linearity and phase distortion

Tonal compression, or nonlinear distortion, is a common example. A camera's dynamic contrast control can be described as a designed distortion. It does not become a fault unless its operation falls outside its specification. It is important to recognise what is a fault and what is not. We will use the literal definition of the word; for our purposes distortion is a fault condition. Figure 9.1 illustrates the point.

Compression is a form of distortion that can arise in a number of situations. Equipment performance deteriorates as ageing components cause the circuitry parameters to change. Considerable effort is taken to ensure circuitry design has sufficient tolerance to combat this but there comes a point when the design is no longer able to cope.

The simple ramp signal is able to reveal this as Figure 9.1 shows. Colour bars will also show the effect but less obviously. Figure 9.2 shows how.

Another clue to the presence of nonlinear distortion is where sync to picture ratio departs from the 70/30 ratio (NTSC uses a slightly different ratio. See Appendix). When the test signal is replaced by a picture, watch

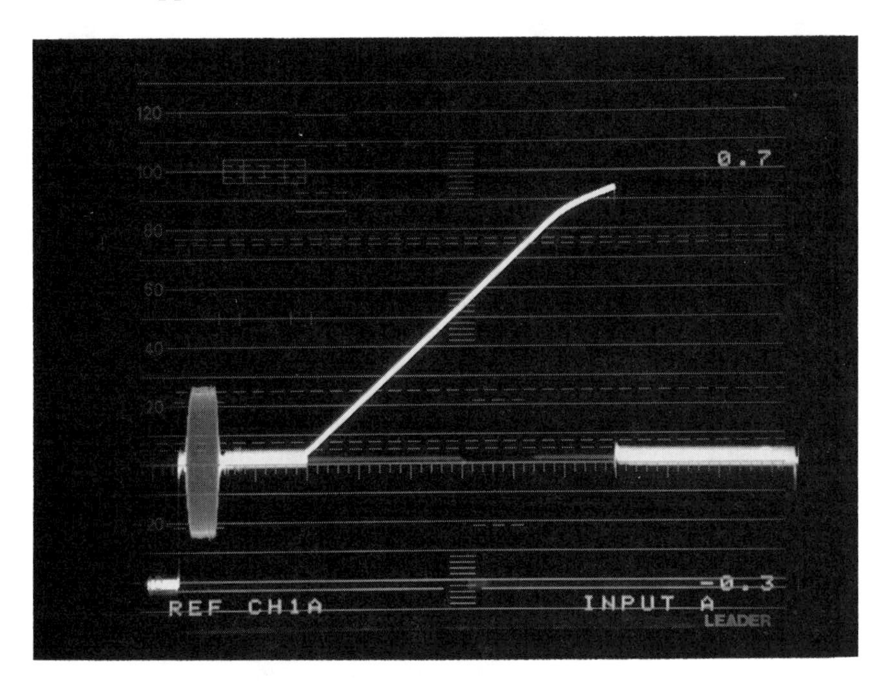

A ramp signal is a straight line, but here it has become curved by circuit nonlinearity. The upper portion of the picture will therefore be compressed. Note the similarity with camera contrast control described in Chapter 6.

Figure 9.1 Nonlinear distortion

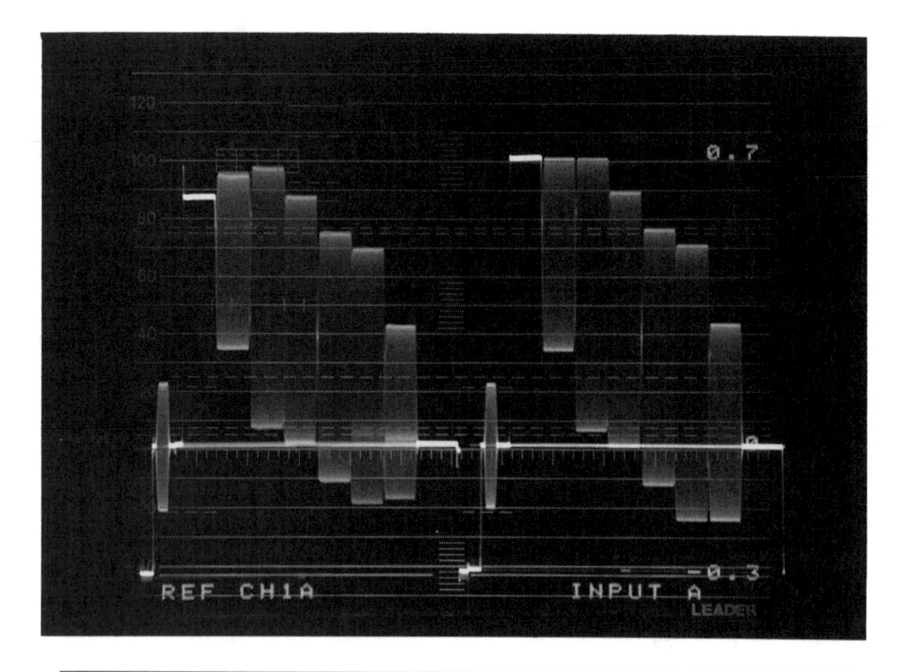

The faulty signal is on the left. The step heights are distorted, but the chrominance makes faults difficult to see. It is only the white bar level that indicates a fault.

Figure 9.2 Colour bars show nonlinear distortion

out for the clip level of peak white, check that it is 0.7 V and that syncs are 0.3 V. If one is right and the other not, there's obviously an error somewhere. But you won't know whether it's incorrect clip level or nonlinear distortion until you run a test. Nonlinear distortion could have occurred *after* camera peak white clippers have done their job.

An example is a distribution amplifier, the device should be transparent allowing the signal to pass through without change. The waveform occupies so much space, it has an amplitude that must fit into the circuitry parameters: power supplies must provide the necessary power. When ageing takes place, circuitry components drift out of tolerance and signal headroom reduces or shifts. It is the signal extremes, syncs or whites, that suffer as a result.

A variation of nonlinear distortion produces another common fault in composite video. Differential phase distortion often arises from the same conditions that cause nonlinear distortion. Yellow has a high luminance value and occupies one extreme of the waveform; it is the first colour to suffer this kind of distortion. Figure 9.2 shows the yellow envelope is no longer symmetrical about its mean luminance value. Figure 9.3 is the vectorscope's view of what has happened and reveals a more serious problem: the fault has caused the yellow bar to have a phase error.

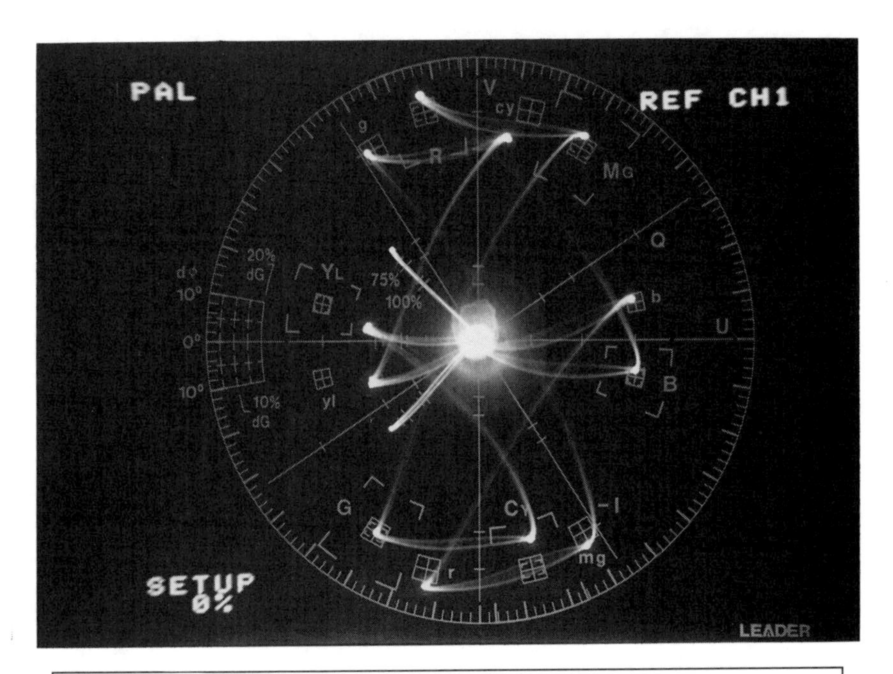

The yellow vector has been skewed off its correct axis towards the green.

Figure 9.3 Differential phase distortion

If you see a phase error on colour bars, check to see if the blue and burst are in the correct phase relationship. Because the blue bar lies nearest to black, a nonlinear distortion is unlikely to affect it as much as yellow. So if blue and burst are in the correct phase relationship and yellow is not, it is a nonlinearity problem. More serious differential phase distortion is where the burst to overall chrominance shifts. It is easy to confuse this with a timing error, as we saw in studio and video tape, and the result on picture is similar. The cause must be checked out with care. As one would expect, the vectorscope illustrates this in Figure 9.4 and appears the same as the earlier examples of timing error.

Figure 9.4 shows a condition that should not arise in composite video. It is not a conventional fault, such as nonlinearity. It is due to picture manipulation in a vision mixer or video effects and is caused by path length difference. The picture has been removed from its black and burst for manipulation – for example, a graphic insertion. Afterwards it is replaced onto the black and burst. The problem arises because the two have passed through different delays and arrived back out of phase. This condition is common where complex mixing and inlaying is carried out in systems that have been expanded and stretched to do more than originally designed for. Path length timing is critical where colour phase is concerned as discussed in Chapter 7.

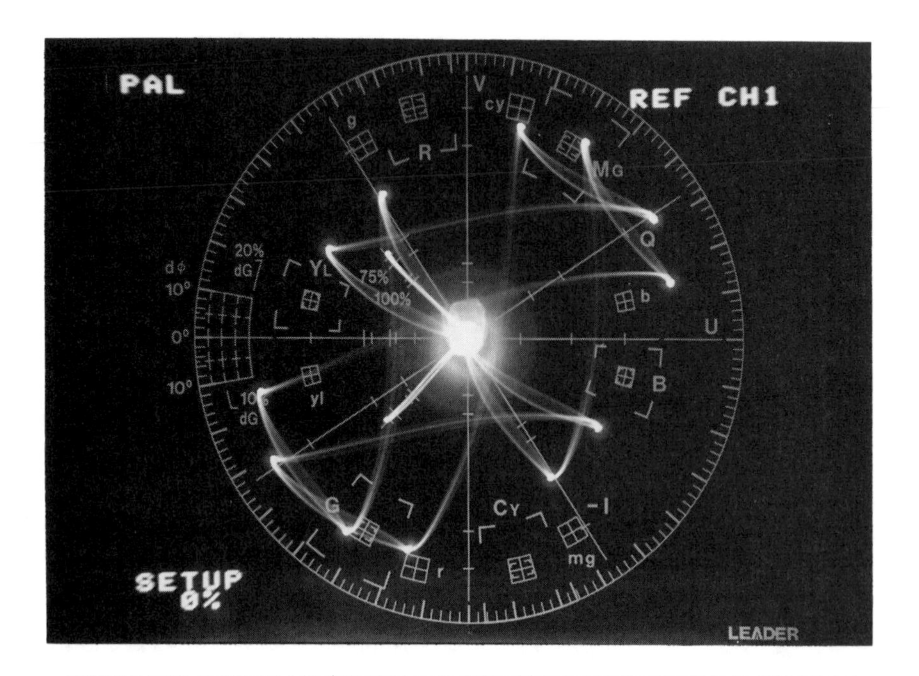

This is just what the principle of composite colour is supposed to avoid: burst and chrominance are generated by the same modulators from the same subcarrier, so how does this happen?

Figure 9.4 Burst-to-picture phase error

DC errors

Video is a DC signal, but such a description does not completely define the waveform; a direct current does not vary and video most certainly does. In fact, the waveform has both DC and AC components: the DC element is black level which is given a fixed value regardless of picture level. Circuit conditions, however, cause variation in the DC level and correction in the form of DC restore is common practice. A common cause of DC error comes from power circuit interference.

Power is distributed as AC (alternating current) and may induce its own waveform into signal circuits under certain circumstances. Once there, it distorts the DC component and becomes part of the signal. The power frequency, 50 Hz in Europe and 60 Hz in America and Japan, is closely related to picture field rate and would therefore appear on picture as a dark band, either stationary or slowly moving. Figure 9.5 shows an example of power interference in a video signal.

Figure 9.5 shows how the signal has been affected by interference. The amplitude from sync to peak white remains the same throughout but the

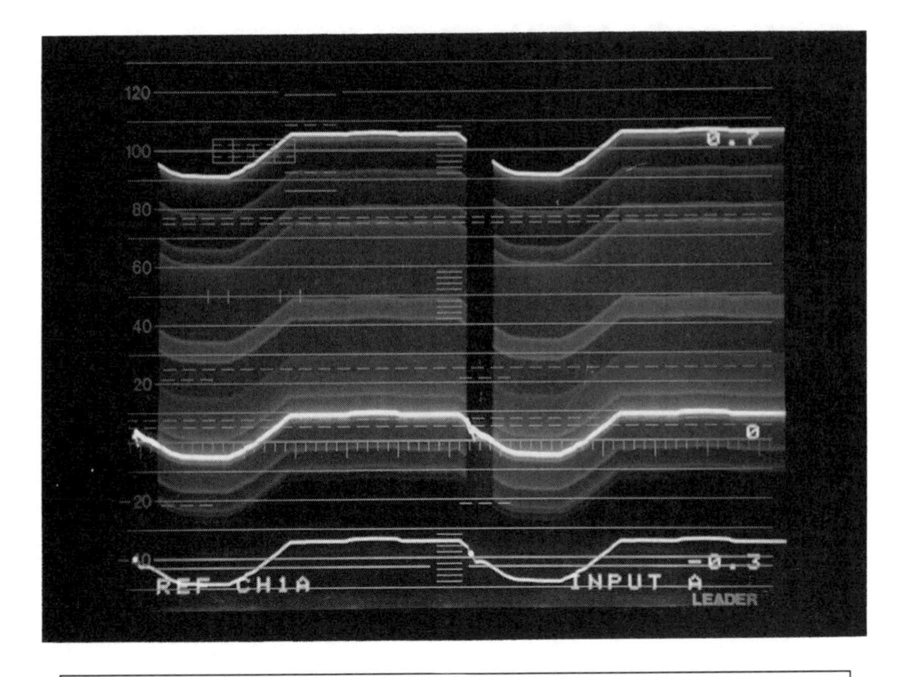

The time-scale is 20 ms. This is the picture frame rate of two fields (2V) and the interfering waveform of the 50 Hz power line is therefore a similar rate.

Figure 9.5 Power interference

absolute voltage varies with the interference. By establishing the back porch of the signal as reference black, we can design our systems to place back porch at its correct level and restore the DC component.

The DC error is not corrected at source; it is removed afterwards. There are not many fault conditions that one can remove after they occur, but this is one. A fault in the signal DC component is correctable because we know what the value should be.

Removing distortion after it has occurred is an uncommon luxury. Unlike DC restoration where black level is constantly placed at zero every line (see Appendix), it usually requires that a measurement is made and a correction added. The difficulty is not in establishing the amount of error but in building a perfect correction. Nonlinear distortion is not easy to put right after the event. Errors that are remedied afterwards are those where defined values are known and can be measured, and accurate corrections made. One example is frequency response.

Frequency distortion

From earlier chapters on signal transmission, we saw how a signal's bandwidth was worked out. Bandwidth affects how quickly a signal can

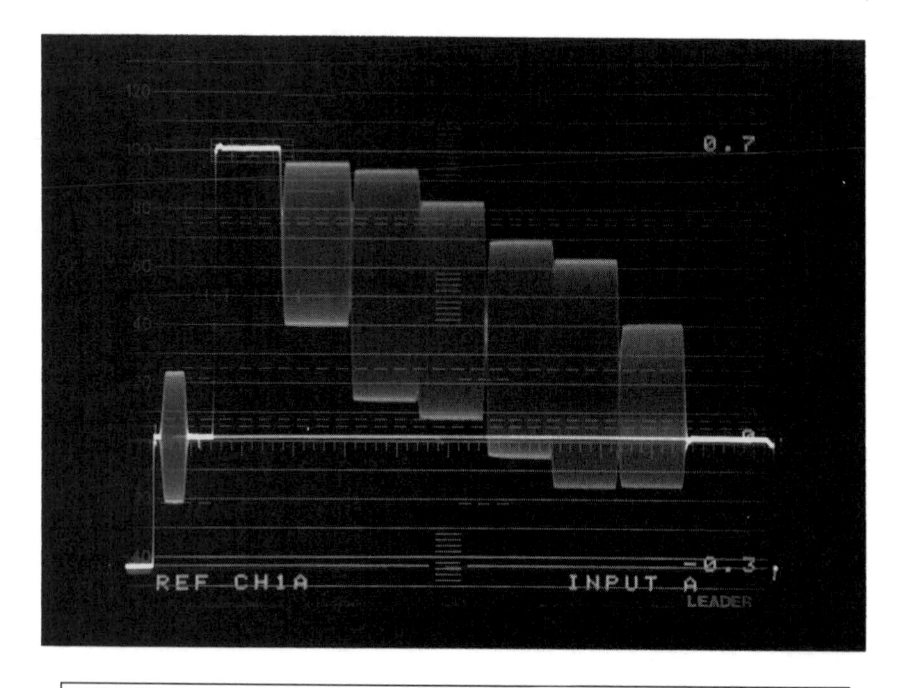

The colour bars show low saturation because of a condition in the circuit causing high frequency loss. Colour subcarrier lies at the upper end of the band and is therefore most affected. Compare with the colour bars in Figure 5.10.

Figure 9.6 High frequency loss

change from black to white, or vice versa. Edge response, or **rise time**, as the term is called, describes this. In Figure 9.6 the colour bars have low chrominance due to high frequency loss in a circuit. High frequency loss and edge response are closely linked.

Standard video has a lower limit of 25 Hz frame rate and an upper limit of 5 MHz. That is the video bandwidth. The limits are normally accepted as being at the 'half power points' – these are the frequency extremes where the signal transmission falls to half power due to circuit losses. High frequency (HF) loss takes place in all circuits, with cable and amplifiers contributing towards the condition. Figure 9.6 illustrates the subcarrier loss resulting from a long length of cable. Other effects also manifest themselves from HF loss. Slowing of edge response shows up in line sync and the rise time becomes longer.

Figures 9.6 and 9.7 both show bandwidth limitation occurring at the high frequency end. The colour bars show reduced chrominance which will mean reduced colour saturation. The voltage pulse in Figure 9.7 shows the shaping due to HF loss. The design intention is for the HF limit to be just high enough for the signal to pass without loss. Excessive bandwidth is wasteful.

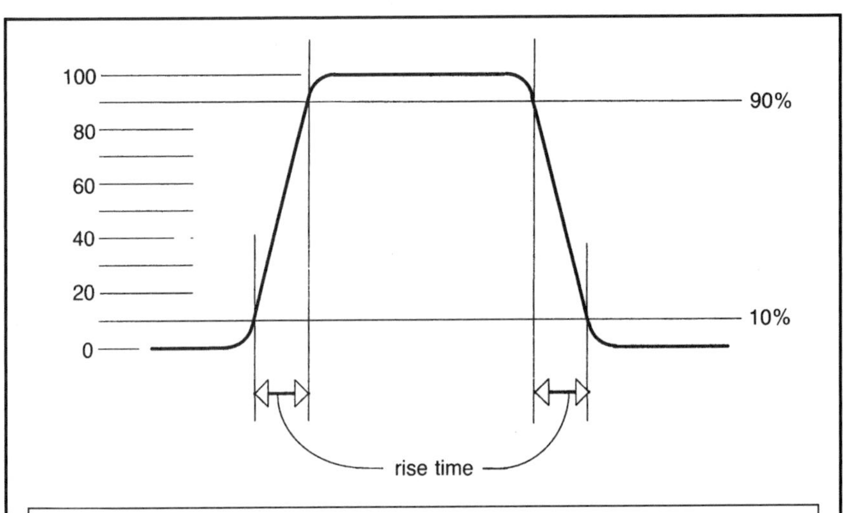

This is a pulse of signal from black level to white. Rise time is measured between 10% and 90% levels, either in the rising or falling transitions. Note that the waveform shown has rounded corners as well. The shape is that produced by a practical circuit having normal bandwidth limitation. The rise time must be sufficient for the signal concerned.

Figure 9.7 How rise time is measured

Pulse and bar

Frequency response and rise time are related by bandwidth, although not all circuits have identical effects. The theory of frequency response is complex and we need only look at its affect on video. A specialised test signal to measure this is the pulse and bar. As its name implies, it has a pulse to show rise time and a half-line bar. See Figure 9.8.

Frequency distortion is not only a loss, there is phase delay to be taken into account as well. This is not to be confused with subcarrier phase, although the two are not unconnected. In building up a step transition, the whole bandwidth of frequencies are required, as described in Chapter 2. All the frequencies involved must start at the same time to produce the ideal step shape. Where they fail to do so is described as phase distortion, this normally accompanies HF loss. How this is shown by the pulse and bar is revealed in Figure 9.9. The distinction between the sort of phase distortion referred to here and chrominance phase is important.

The pulse and bar method measures by comparing waveform shape. The pulse is accurately formed and frequency distortion changes that shape in a specific way. Complex measurement is now not required to describe a circuit frequency response; it is sufficient to state what is

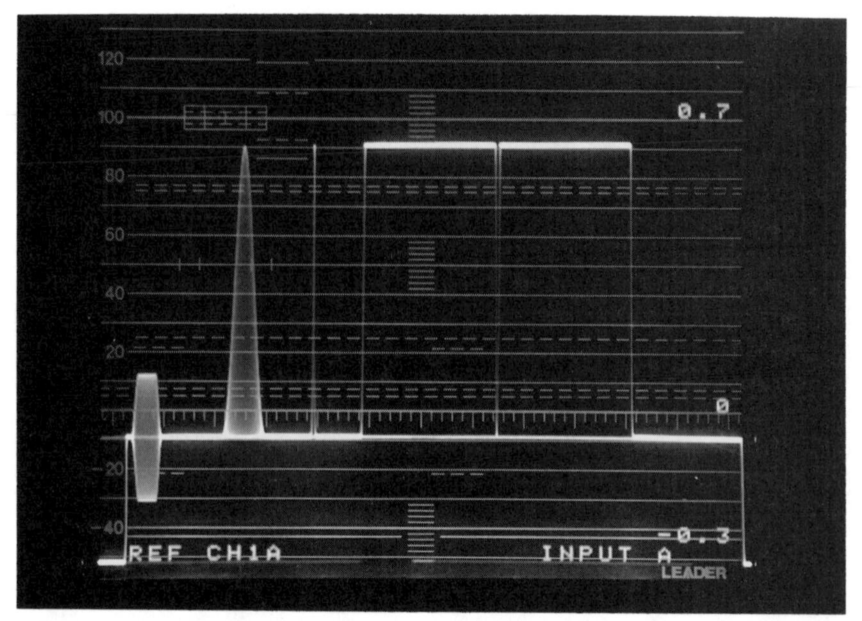

(a)

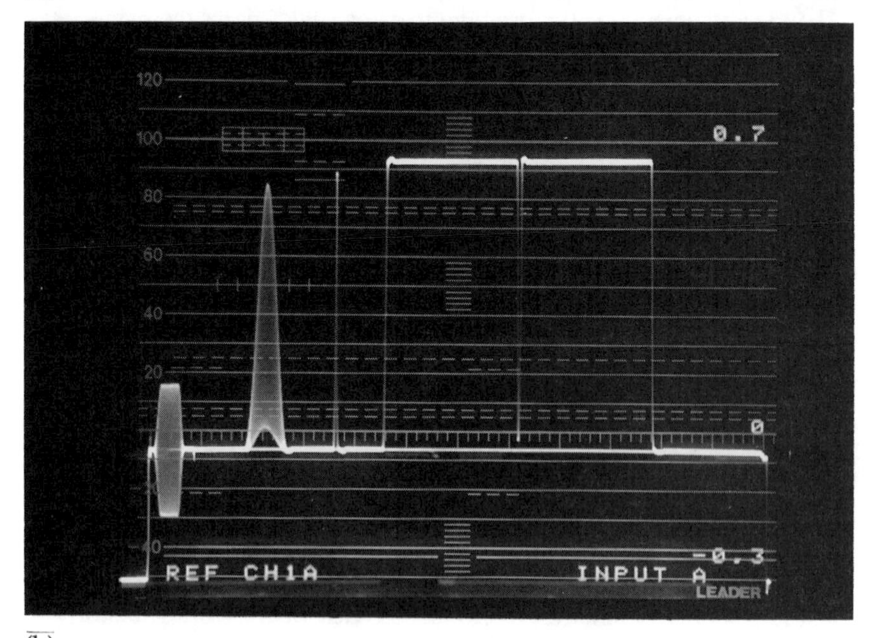

(b)

Pulse and bar checks frequency response. The pulse is a sine^2 function meaning it has a sine wave form. Loss of HF causes the chrominance pulse at left-hand of the waveform to reduce in height. The original is shown in (a) and the distortion in (b).

Figure 9.8 Pulse and bar

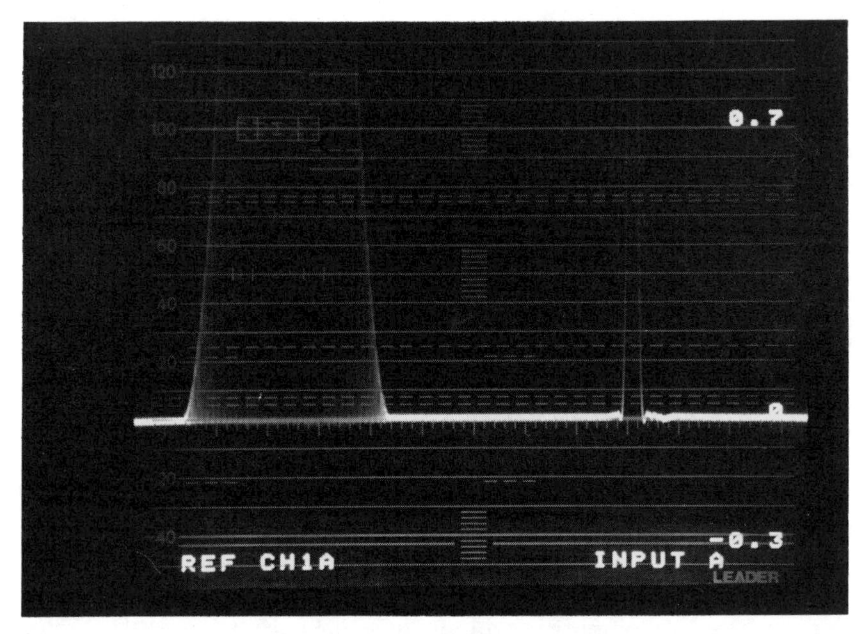

(a)

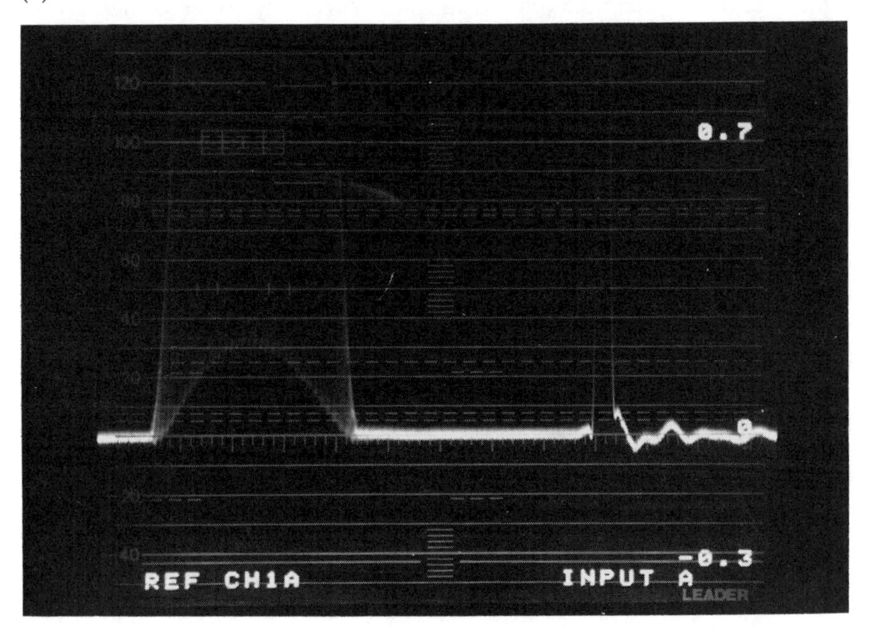

(b)

Magnifying the display in Figure 9.8 reveals the HF loss clearly and also shows, in the luminance pulse on the right, shape distortion due to phase delay at high frequencies.

Figure 9.9 Magnified pulse shows HF distortion

known as the **k rating**. The standard shape, and various amounts of distortion, may be engraved onto a WFM graticule for the purpose.

The bar is likewise a predetermined shape. Its height is 100% and it is there to show up waveform low frequency loss, or sag. LF loss is another circuit limitation and, although video has a DC component (where frequency = zero), practical systems are not designed to pass DC – a shortcoming remedied by DC restoration, as we have seen. From this we can see that the earlier reference to 25 Hz as the lowest frequency need not be adhered to. It is quite reasonable to say that the circuits need only pass line frequency and above and let DC restoration at the destination make good errors of black level.

The amount of LF loss can now be measured as a line sag and defined by the distortion of the bar. Again, the degree of distortion is expressed in the form of a k rating.

Other test signals

There are a number of test signal generators available offering a range of signals; many are composite/component switchable. It is as well to investigate the actual facilities offered to make sure they cover your requirements. The NTSC/PAL differences must be considered when obtaining

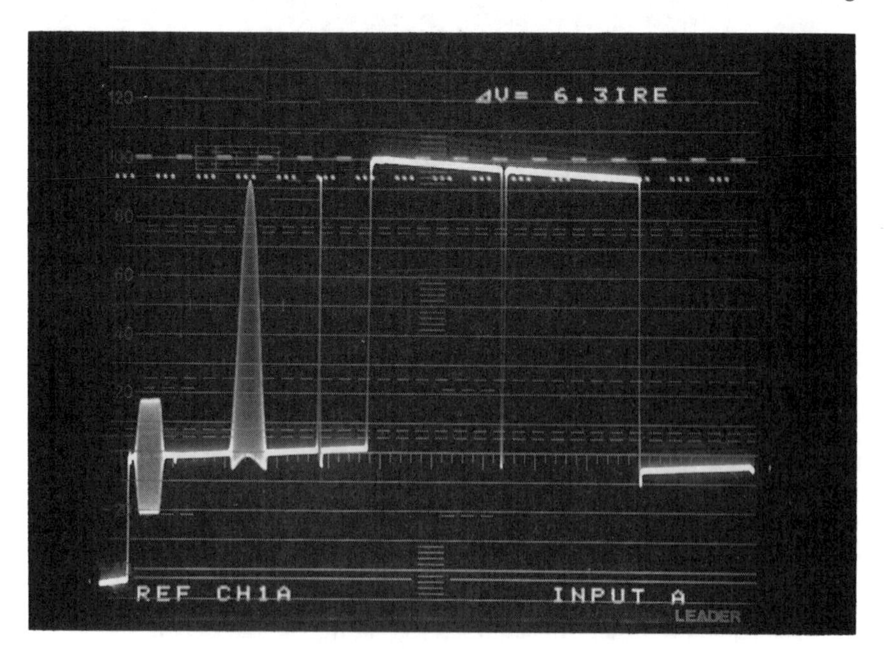

How much sag the bar has indicates LF loss. The cursors show this in percentage (shown in IRE values).

Figure 9.10 The bar measures LF loss

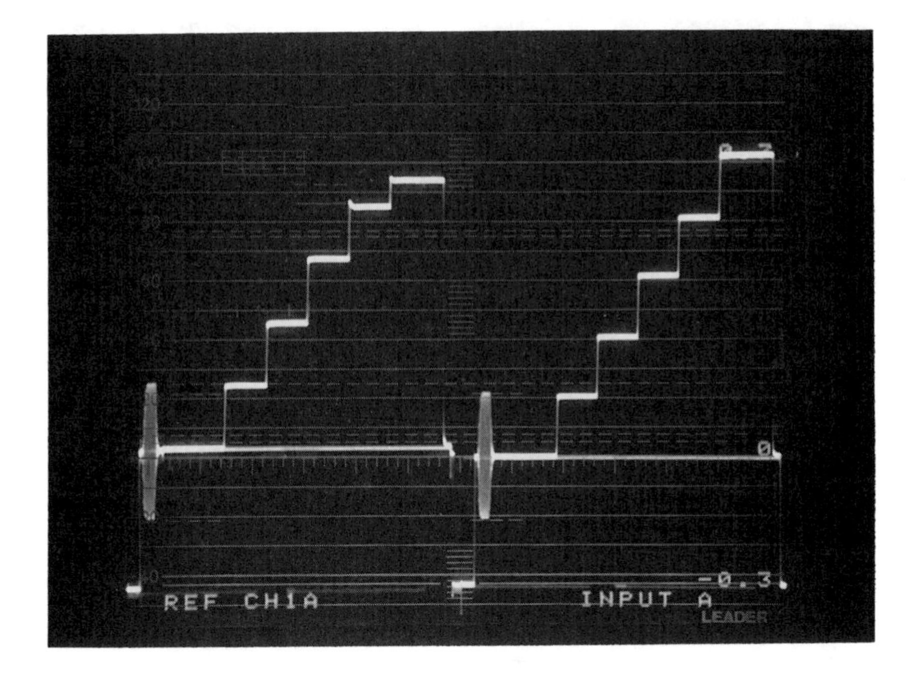

The step is for nonlinearity measurement. The equal step heights will vary if nonlinearity is present. Compare with the ramp test signal used in previous tests.

Figure 9.11 The step

equipment outside the country of use to make sure they comply with local standards. Also, there are differences in component video standards used in the various recording standards and these, too, must be borne in mind. Use only test equipment that is specifically designed for the standard concerned.

All video test signals have one thing in common; identical sync and burst. It would completely defeat the object of test and measurement if this important part of the signal were to be distorted. In each case the signal must enter the system as perfectly as possible and must not cause parameters to alter by any variation in the synchronising components of the signal. Three examples are described in Figures 9.11, 9.12 and 9.13.

Calibration

All test signals are calibrated signals, and colour bars are the universal calibrated test signal, by definition. WFM's and vectorscopes have calibration checks. These are internal signals used to check the accuracy of the equipment, a kind of self-check. Calibrated cursors, on the other

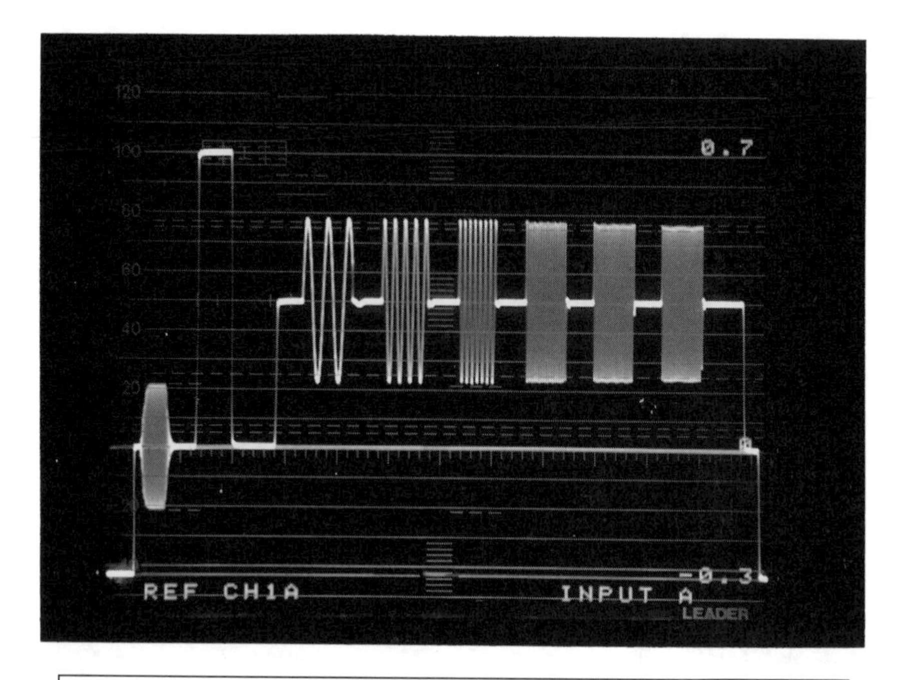

Multiburst is a group of frequency bursts. As the signal passes through a circuit, the bursts' amplitudes will alter if frequency distortion is present.

Figure 9.12 Multiburst

hand, are adjustable markers that are placed against the signal, giving the value as a read-out. Such an example is shown in Figure 9.9 where the sag of the bar is measured by the cursor positions.

A calibrated signal may be a simple voltage switch as in Figure 9.14.

Comparing signal to signal

So far we have measured our signals by comparing them to a scale or calibrated waveform. An alternative is to compare like with like, that is, to a standard signal, which is a chosen signal that has been set up accurately. The important point here is that the chosen signal is now a standard, or master, and must be guaranteed. Taking colour bars, we can set these up and display on WFM and vectorscope. Now overlay other signals in turn and see how they differ from the master.

Overlaying to compare has been demonstrated already in these pages, but is not always ideal, for one signal may easily obscure the other. A common method is to switch between them showing them alternately. Not all test equipment provides the option but it is quite possible to manually select signals in turn and is a particularly good way to see quite small differences. One useful feature is to make the selector operate

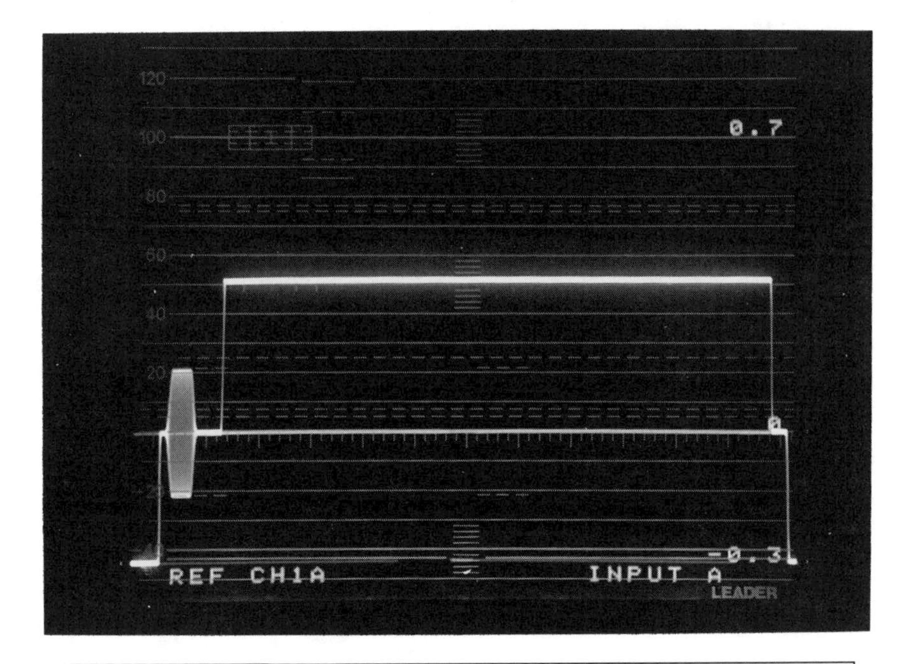

Flat field is a uniform level over the complete line and field. Shown here as 100% flat field, it is used to determine the low frequency performance of a circuit. The flatness of the signal should remain constant from the top of the field to the bottom. Also a demanding picture display test indicating purity of grey across the screen.

Figure 9.13 Flat field

electronically in the field sync period. If all the sources are in sync, the changeover will take place in the next field blanking period. The cut from one source to another will be clean with the display remaining steady. The switch can be rapid or as slow as required; as long as it makes signal differences clear, it really does not matter.

Switch speeds as fast as line rate can produce; what is known as a 'parade'. Each successive line shown by the WFM will be the next source, and so on, one after the other.

Comparing signal to signal is primarily to get matched signals. As a measurement method it is quick and accurate but does require that the master is accurately calibrated. On the other hand, any specific features of your system can be replicated by setting up the master as you require. In setting up cameras in the studio the final trimming of black, exposure and colour is done on a picture monitor; never overlook the importance of using the picture. Just make sure the master camera is correct. A good display is important, whether waveform, vector or picture. There is no point using this method if the display fails to show the differences, for whatever reason.

The subject of test equipment is extensive and only a brief insight can be given here. Test equipment manufacturers' literature is always useful, often giving valuable advice and background information to video measurement.

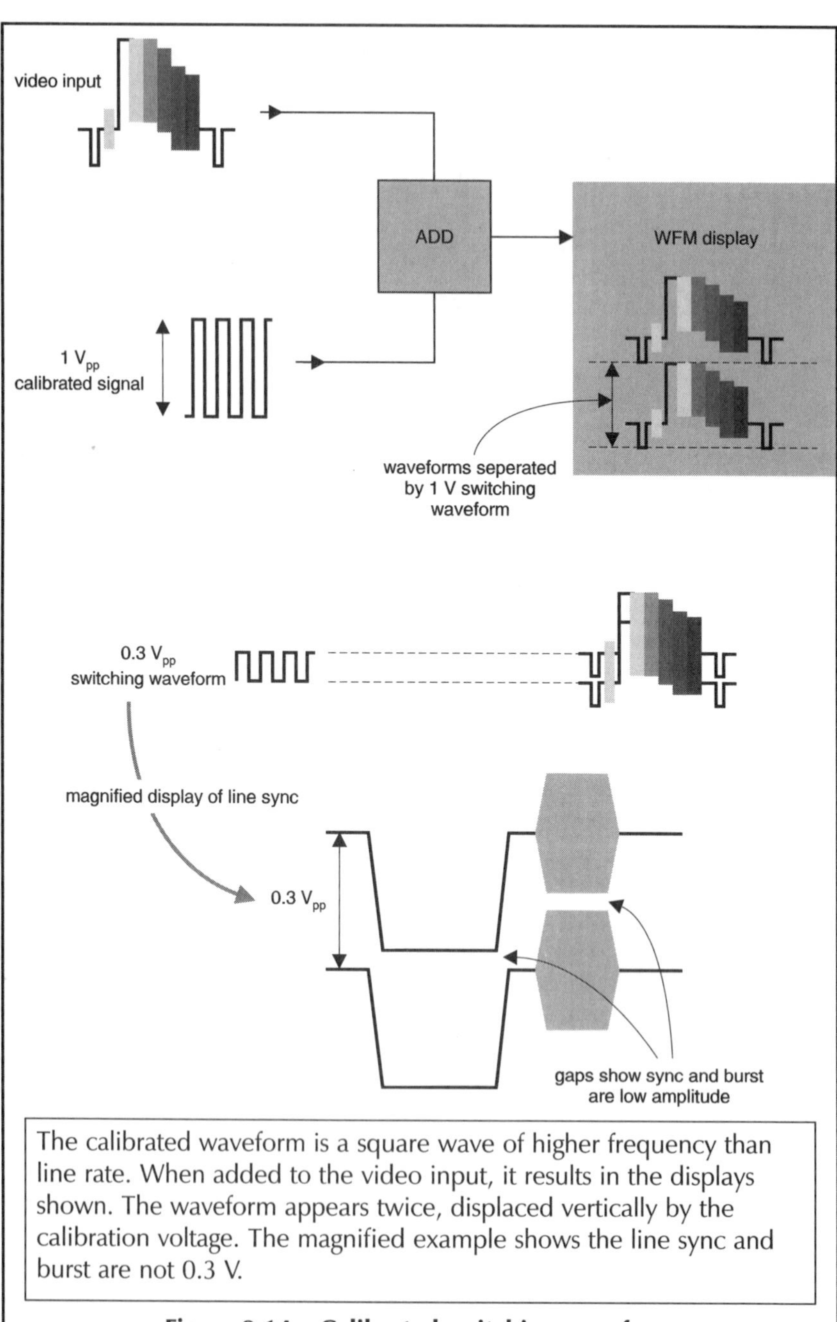

The calibrated waveform is a square wave of higher frequency than line rate. When added to the video input, it results in the displays shown. The waveform appears twice, displaced vertically by the calibration voltage. The magnified example shows the line sync and burst are not 0.3 V.

Figure 9.14 Calibrated switching waveform

CHAPTER 10

ALTERNATIVE MEASUREMENT METHODS

Using the picture monitor

We have already seen how the picture monitor can be used in checking H and C timing. It is the picture that holds the final answer and we must learn to use it properly. Picture set-up, as detailed in the Appendix is critical; the camera set-up in Chapter 6 illustrates how the video parameters interact to confuse the unwary.

Measurement by eye fails because the eye has unreliable long-term memory but when pictures are shown in succession the eye will reveal colour and luminosity differences with ease, particularly the experienced eye. Final camera checks must always be on the picture monitor. Compare the results of setting up sources in the studio described in Chapter 7: the cameras will be white balanced, then, a final check on picture to see how accurate this has been. The picture will quickly show how well the cameras match.

Even more critical is to point all the cameras at a person and study flesh tone differences camera to camera, a technique beyond the basic test equipment described in this book but well within the ability of the eye.

To exploit this characteristic of the human eye, a new measurement technique has been developed. By placing into the picture a calibrated patch, or bar, a reference is created against which the eye is able to compare. The eye's long-term memory problem is now overcome allowing its acuity in comparison to be utilised. The calibrated bar is inserted into the source video signal and appears to the monitor and the eye as an integral element of the picture. This is the Vical system.

The Vical system

Designed as a portable unit, the inserted bar is true grey with variable level. To check camera white balance adjust the grey bar to match that of the white card. The instrument has the facility to remove colour to make the luminance match possible, then by restoring the colour, any remaining colour balance error is revealed. Likewise, a black balance check can also be carried out. Camera pedestal may be measured by the same method.

Vical is a picture-based system. It is not designed as an engineering tool in the same way as one based on waveform measurement. It will only function as a picture measurement device, it cannot measure syncs or timing. But, throughout this book, stress has been placed on the importance of pictures and their relationship to what is seen on the test set. The Vical in-picture measurement method offers an alternative to using waveforms.

Waveforms are convenient and make measurement straightforward but they must be properly understood, one of the main purposes behind this book. But too much waveform reliance in an operational sense can lead to picture blindness, where the individual fails to trust their own eyes and turns therefore, to the WFM for decisions of a pictorial nature. This is not what the waveform is for.

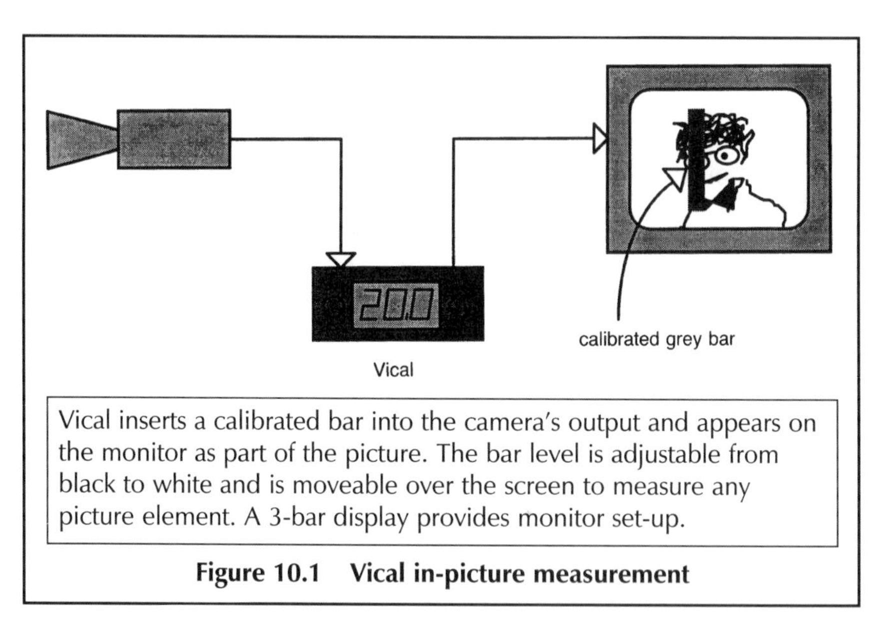

Vical inserts a calibrated bar into the camera's output and appears on the monitor as part of the picture. The bar level is adjustable from black to white and is moveable over the screen to measure any picture element. A 3-bar display provides monitor set-up.

Figure 10.1 Vical in-picture measurement

The in-picture waveform

An alternative to the traditional WFM is to turn the picture monitor into a waveform display. Modern digital processing makes possible signal analysis, then re-processing to present it to the picture monitor as a picture of a waveform. Such instruments have many of the facilities of conventional WFM's and at the same time offer a size and cost advantage, the display being, of course, already provided. An example is shown in Figure 10.3.

Portable instrumentation is becoming more available as microelectronics continue to offer more and more facilities. The essential requirement never varies: accuracy and reliability – there is little point in using a ruler with 'wrong inches'. Give plenty of thought to your needs, choose carefully and buy wisely.

Here Vical is used as a spot meter. The level of the grey bar is adjusted to match the skin tone. The read-out shows what level the bar has been adjusted to. Therefore, in this example, the skin tone measures almost 60%. Compare with Figure 10.3.

Figure 10.2 Measuring with Vical

The advantage of in-picture waveform and vector displays is that the display size may be as large as the picture, or, as here, waveform and vectors shown side by side. Showing the picture at the same time is also most useful. Versatile waveform and vector monitoring in a compact battery-operated instrument is of particular value in location and remote situations. Using the waveforms and vectors display for actual pictures is of limited value. However, the fact that here the signal has exceeded 100% may be very important.

Figure 10.3 In-picture waveform

CHAPTER 11

INTRODUCING DIGITAL VIDEO

Digital video is a development of the systems already described and is based on the conversion to a digital form of either component and composite. When the video signal is in digital form it is highly manipulative; it is also very unlikely to be affected by common distortions.

What is digital?

The conventional way of making an electric signal is to use the parameter of amplitude: this is the analogue signal that has been the subject of this book so far. We have seen how the analogue signal is vulnerable to amplitude nonlinearity distortion and frequency response limitations. The digital form does not have an amplitude parameter, at least, not one that is as sensitive to distortion as its analogue counterpart.

Converting from analogue to digital means that a standard video signal, component or composite, is subject to amplitude measurement at a pixel rate, and the value of each pixel converted to a number. The critical part of the digital system is the conversion process. When analogue to digital conversion takes place the video signal is sampled to measure its level, a process that must be done at a rate at least twice that of the highest frequency of the signal. In the case of composite, the sampling rate, as it is known, must be twice subcarrier frequency if the colour is to be analysed correctly. Going back to bandwidth, if the signal's highest frequency is 5 MHz, the sampling frequency must be at least 10 MHz. meaning that the electronics operate twice as fast as in conventional video.

Binary is two-state and, in electrical terms, a circuit is either on or off, which translates directly from the binary 0 or 1. We discussed two-state electric signals in Chapter 1 and saw how simple they were. Only now the speed has increased by ... Well, suffice to say, sending a code describing any one of one hundred levels every one ten-millionth of a second is very, very fast compared to the methods described then. But the essence of simplicity remains.

Conversion works like this. The most significant bit is to determine whether the signal exceeds 50% or not. If it does, next determine whether it exceeds 75% – that's the next bit. If it does not reach 50%, does it get to 25%? And so on. This principle of breaking every part of the signal into bits and giving a number to the bit series is the basis of digital signals.

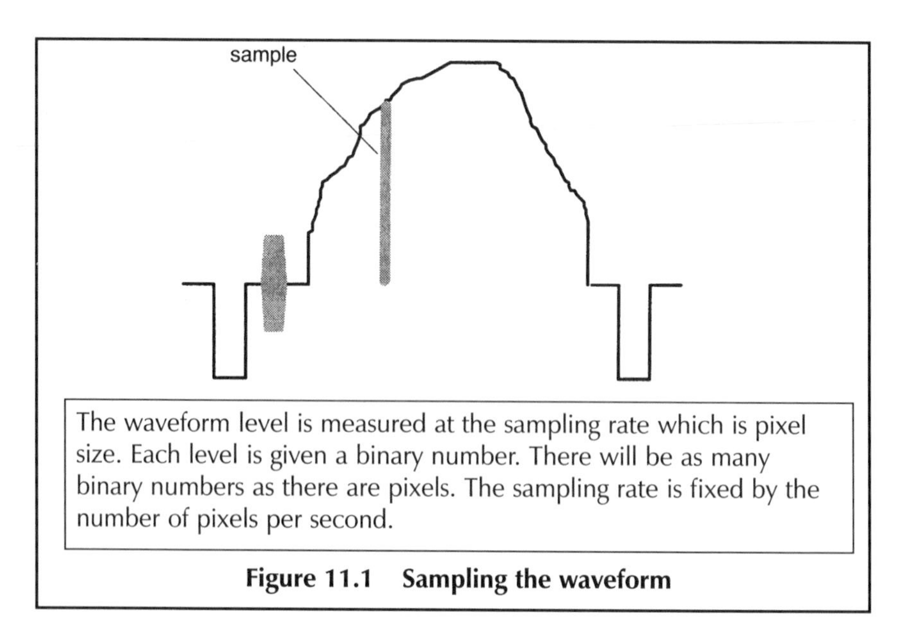

The waveform level is measured at the sampling rate which is pixel size. Each level is given a binary number. There will be as many binary numbers as there are pixels. The sampling rate is fixed by the number of pixels per second.

Figure 11.1 Sampling the waveform

Binary numbers are to the base 2, that is they only count up to two, the numbers are zero and one, hence the 0's and 1's used in digital language. A simple binary series that will count up to 8, or take 8 samples, requires three circuits and is shown in Table 11.1. Eight samples means that the system can distinguish eight levels of signal.

Table 11.1 Binary series

Sample no.	Circuit A	Circuit B	Circuit C
1	0	0	0
2	0	0	1
3	0	1	0
4	0	1	1
5	1	1	0
6	1	0	1
7	1	1	0
8	1	1	1

Three circuits are required to carry this amount of data. Circuit A carries the most significant bit, Circuit B the next, Circuit C carries the least significant bit. Loss of the circuit carrying the least significant bit is the least disruptive but loss of Circuit A would be catastrophic.

Every level is given a binary number and if we require 100 levels to describe our signal (based on the accuracy of 0 to 100%), a count to one hundred will take seven circuits. In fact, 100 levels are inconvenient; if we do the binary arithmetic, we find the number comes out at 128. Digital video therefore requires at least 8 bits.

When first describing the video signal and how it is transmitted, one option was to send each pixel along its own circuit. An impracticable option because of the circuit complexity. Here we have a similar situation with each data bit allocated a circuit of its own. This is termed **parallel data** and is shown in Figure 11.2.

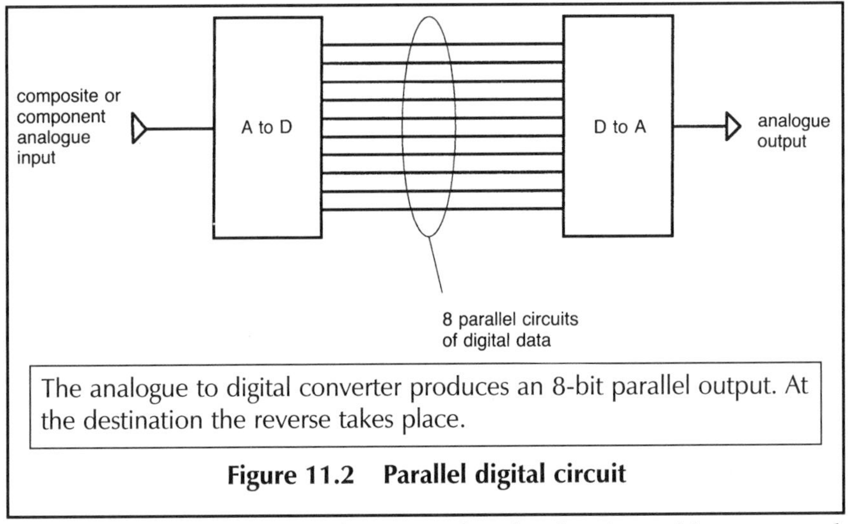

The analogue to digital converter produces an 8-bit parallel output. At the destination the reverse takes place.

Figure 11.2 Parallel digital circuit

A more convenient method is to serialise the data into a bit stream and send down a single circuit as in Figure 11.3. The penalty is that the data rate has to increase in order to transfer the same amount of data in the same time-frame.

The data rate will depend on how fast the signal can be sampled and what level of accuracy is required. Going back once more, telegraphy used a data rate that worked at human operator speed, fitting the cable technology of the day. If we want to send digital video at this speed we could do so but would have to reduce the frame rate so much that the signal would appear as a sequence of stills, probably no faster than one a week. Alternatively, we could trade picture quality for speed.

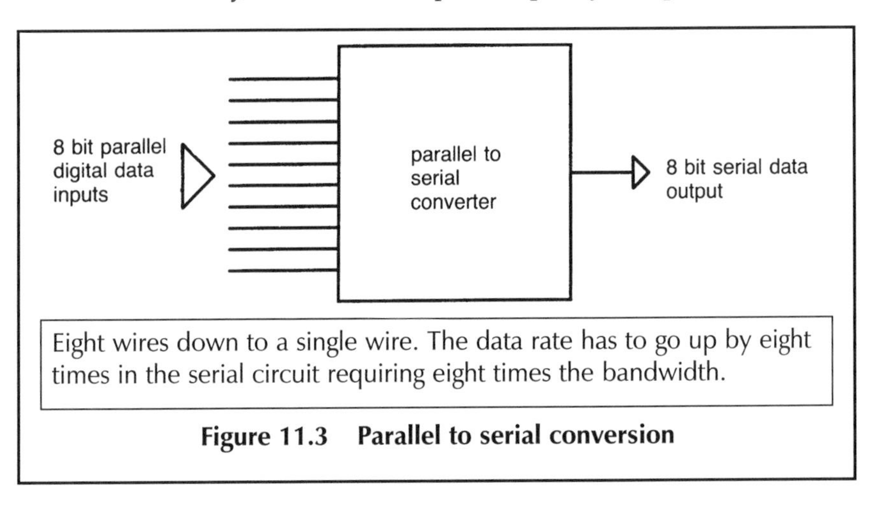

Eight wires down to a single wire. The data rate has to go up by eight times in the serial circuit requiring eight times the bandwidth.

Figure 11.3 Parallel to serial conversion

Packaging video into a digital form demands that another standard be created. Standard CCIR 601 is based on component video, Y, U and V, sometimes seen written in the form Y, Cr and Cb. To get the necessary speed of transmission the sampling rate is 13.5 MHz and requires a circuit bandwidth of between 200 MHz. and 300 MHz. for the serial data rate. So we begin to see how much more demanding digital video can be of circuit bandwidth.

But it is the act of conversion that is most demanding. The conversions, analogue to digital and back again, are the points most vulnerable to error. Measuring discrete levels means that approximations are made. Take those near to black. Distinguishing between 0 and 1% is a far more significant level change than distinguishing between 97% and 100%. Yet the digitising system will give equal emphasis to both.

The effect is called quantising and is reduced by increasing the number of bits. A one-bit system can only distinguish two levels and the quantising distortion will be very high indeed. At very low level signals around black level, quantising distortion is seen as discrete tones rather than the gradation of the original analogue signal. It is a matter of deciding how much tonal resolution is required for a practical system.

From this it would seem that the action of digitising a signal and back again, will inevitably distort it. Well, it does. Digital signals are not free from errors. So it is highly desirable to avoid needless conversion and reconversion. Once converted to digital, as early as possible, it should remain in the digital domain as long as possible. Ideally, right up to the very end of the chain at the picture display.

Video is more forgiving than audio in this respect because the number of levels, or bits, required is not so great. The audio dynamic range is many times greater than the video contrast range. As with any system, care must always be exercised over the design and usage of equipment if we are to benefit fully from the advantages of digital video production.

Whereas analogue will, if the signs of signal degradation are read properly, give warning of trouble, digital will not. By its very construction it ignores one of the principal analogue parameters; amplitude – until it's too late. When the digital destination is unable to make sense of the signal because, for example, it has become lost in circuit noise, the result is total collapse. But that is very much further than the noise tolerance of analogue.

How digital video affects operations

Why go digital? There is one overriding reason: when in the digital domain, the signal is extremely well protected from distortion or interference. The digital signal is no more than 0's and 1's, two levels only and, as long as the detection system can distinguish a 0 from a 1, the system will function. Video tape generation loss in editing is no longer a problem because the machine now works only in two levels, not the one hundred levels, plus chrominance, as in analogue.

Cables and circuits no longer need to operate from DC to the highest frequency. Interference from power cables, noise amplifiers, all will be rejected or ignored.

Timing and synchronisation become part of the digital operation, the synchroniser is a digital memory that easily fits into a total digital system. Manipulation is another established digital process that is easily integrated. When the signal is digitised early on and all processing carried out in the digital domain, the signal does not undergo unnecessary conversion or re-conversion.

Wide screen video is more easily accommodated into the digital domain than in analogue. The wider screen demands an increase in horizontal resolution. The scan rate remains as before but as the line is longer, it has more pixels, the information it contains is higher. This translates directly into a data rate increase. The whole concept of wide screen is based on a better picture. Digital processing reduces degradation and so fits well into this concept.

Measurement of digital video

Placing the signal into digital form removes a lot of measuring requirement but does not preclude the need to monitor. We would find it impossible to look at a digital video circuit with a standard WFM and make sense of what we see. A binary number as a waveform is meaningless to the eye and watching a series of pulses for every pixel level at pixel rate, is just not practical.

Automatic monitoring of the data stream with equipment that is able to analyse and indicate the state of the signal is the only way. The whole operation can be in digital form – there is no need to convert to analogue – the monitor simply looks at the data. The result is a read-out of the condition of the circuit in a form the operator is able to understand.

Should the digital signal become corrupted, for instance, a bit goes missing, then the system should be self-checking and compensation put in place. In a practical system, it is quite reasonable to expect a bit to get lost every so often because of noise or interference. Video tape dropout is a good example and an automatic monitor that can spot the problem can also make a good guess at what the bit should have been. It has only to look at the bits either side and interpolate to get the answer.

Here we are tied very much to how the signal monitor functions, how it evaluates and what emphasis it places on signal parameters. But with digital video there is little option but to rely on such equipment. Digital systems, whether signal processing or monitoring, are on the whole, reliable – a feature that directly stems from the two-state binary operation. If this were not so the increased complexity of digital video would mean a very much more unreliable system.

A digital signal is vulnerable in one aspect. As electrical noise increases it may interfere with how well the 0 to 1 transitions are read. To read the signal, the digital to analogue converter must know what part of the signal is which and to do this it must be able to synchronise itself to the incoming data stream. If it fails to hold synchronism because of noise interference, corruption occurs. The conversion back to analogue should ideally be as close to the end of the chain as possible. But by the time this takes place there will have been considerable processing and routing over various circuits on the way. Every stage is a potential source of noise and corruption but with good design and proper use digital processing is able to offer much greater flexibility.

Compression

The processing power available to the digital signal far outstrips that of any earlier system. It is now quite feasible to assess signals for redundancy, or how much of the signal can be discarded whilst still retaining its viability. Moving pictures are not all movement; for instance, there may be a still background behind a presenter. If these areas of still image are not being constantly transmitted, a saving of circuit space and usage can be gained. This is an example of signal compression. Gains can always be made by trimming away redundancy and optimising the system at all times.

Compression will, at some point, degrade the signal, particularly where the picture has considerable movement. Such procedures have to be carefully evaluated before incorporation into the standard; different amounts of compression apply in different circumstances. The MPEG committee is responsible for assessing the effects of compression and makes recommendations accordingly.

The quest for quality

Great emphasis is placed on the quality of digital signals. In reality, these are only as good as the source material and it is fortunate that signal acquisition has made significant advances as well. Where digitised signals score so dramatically is because of the immunity to signal recording and processing distortion that has so plagued analogue. As long as the system is correctly designed, installed and maintained to its specification it will always appear transparent.

Digital cameras offer predictable parameters accessible via software. The fear of moving a control is removed because we can simply ask the system to restore to the original value. This is a great leap forward, the opportunity to make pictures just as we want. It does not, however, remove the photographic prerequisite to understand what these parameters mean.

Because of the extensive networks based on PAL and NTSC there are many proposals put forward to adapt these to digital or higher standards of analogue. The interface is the receiver, which is still analogue and unlikely to change significantly in the short term whilst the demand for compatibility still exists. So digital distribution must at some point interface with existing analogue. PAL Plus is an example of digital processing up to the transmitter and then, by means of sophisticated PAL encoding to eliminate, as far as is possible, composite degradation. This provides a more satisfactory approach than conventional encoding at each picture source with the attendant timing and quality constraints that brings.

Digital video standard 601 requires quite different circuit standards to analogue video. The higher operating frequency means different cable design and transmission equipment. Composite video has quality compromises, mainly from the interaction of subcarrier with fine picture detail, a condition made more critical by the PAL system. But very extensive networks based on composite video already exist. Component video has received a great deal of attention because of its inherent higher quality, particularly with the advent of wide screen TV. A new BBC development in the UK allows high quality component video to be sent through standard composite circuits and equipment.

These developments pave the way for the expansion of wide screen video based on the 16:9 aspect ratio. They may be seen as a stepping stone to a fully digitised network.

Finally...

At the end of the chain we still view an image that is analogue. The picture has still to be understood in its own right and the picture monitor will continue to serve its function as part of the assessment process. Pictures are a human creation and technology, whether analogue, digital or any other form yet to be invented, must not be permitted to take control over them.

The business of broadcasting is experiencing a proliferation of standards and offshoots of standards. A situation brought about by protectionism and commercial interests that does little to encourage the sharing of technology. A lot of development is not done for the consumer but to feed other interests. At the bottom line, the standard engineered so carefully is coming under threat for it is seen by some as a hindrance.

The overall theme of this book has been the maintenance of the standard. And the standard, after all, is no more than a mere tool used in the maintenance of quality. Understanding the *whole* system is vital if quality is not to be subjugated by these other interests.

APPENDIX

1. Picture monitor set-up

Correct picture set-up is essential. There is no greater cause of confusion in video operations than that caused by incorrect monitor set-up. Different monitors will give different pictures and where monitors are working side by side it is particularly important that they match each other. Avoid mixing manufacturers and designs wherever possible.

Monitor set-up is a critical operation requiring a known signal source, and should ideally be carried out by one person following exactly the same procedure.

Pluge

The most common test signal is PLUGE (picture line-up and generating equipment) which has a step waveform with additional 5% steps above and below video black level. See figure A.1.

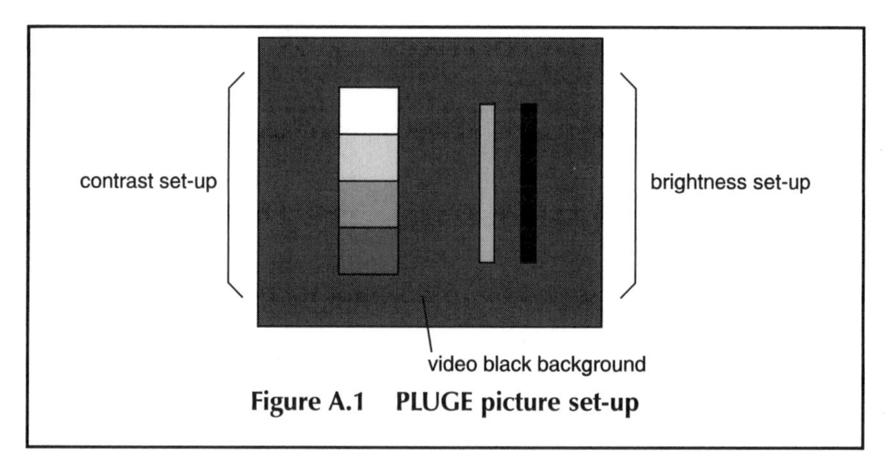

Figure A.1 PLUGE picture set-up

The four-step contrast set-up may be measured and compared to picture display manufacturer's specification. It may also be used to compare monitor-to-monitor contrast and gamma.

Setting brightness is critical to 1%. Brightness is adjusted until the darker of the two 5% bars is just visible. Alternatively, setting the lighter of the 5% bars to be just visible gives a lower brightness setting, and is appropriate where fixed set-up or pedestal is in use.

Vical

Vical picture set-up is part of the Vical measurement system. The test signal is inserted into the source pictures and therefore appear to the picture display as integral with the source signal as explained in Chapter 9. The background is therefore the source picture which is camera pedestal in the case of a capped camera. If camera pedestal is set at 3%, this will show against the black bar which is at video black of 0%. Brightness may be set to just see this difference.

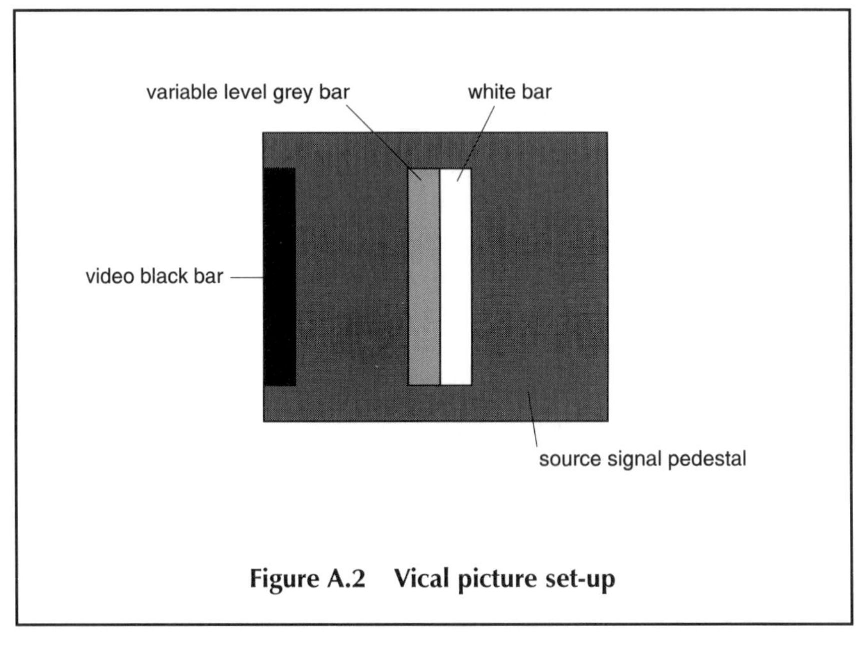

Figure A.2 Vical picture set-up

A more accurate method is to set the grey bar at 1% and adjust brightness to show this difference. Any value of grey bar level may be chosen making the Vical method very accurate and repeatable. It also allows set-up in a much wider variation of ambient lighting conditions.

Current monitor design practice places emphasis on menu set-up. But beneficial as this may be, it does not preclude the requirement to *know* that the picture is properly set up for the viewing conditions in use.

Colour

Measuring the light emitted by the CRT phosphors (or whatever method the display uses) is the most accurate method of setting screen white balance. Proprietary methods are available.

As with the camera, black balance is carried out first followed by white balance. But for the procedure relevant to a particular display, follow the manufacturers instructions.

2. Gamma

Gamma (γ) in television is a product of the cathode ray picture tube. The CRT is based on the triode thermionic tube and this device has an inherent nonlinearity.

Picture tube gamma causes compression at low signal inputs and expansion at high signal inputs, which results in the exaggeration of tonal difference dark to light. The relationship of normalised input to output is:

$$\text{Input}^{\gamma} = \text{Output} \tag{1}$$

For a CRT, the gamma is 2.2. This figure is fairly consistent but some variation occurs between tubes, mainly from design differences and ageing.

$$\text{Input}^{2.2} = \text{Output} \tag{2}$$

Equation (2) is the transfer characteristic of a standard picture tube set out in Figure A.3. A gamma of unity is shown as a dashed line for comparison.

Black-and-white television benefited from the tonal compression near black as the effect of noise was subjectively made less objectionable to the viewer. Gamma compression was not acceptable for colour television because colour saturation became distorted, and the opportunity was taken to design a linear system. To correct a gamma of 2.2 requires the opposite gamma of 0.4545.

Good noise performance was still considered necessary and the correc-

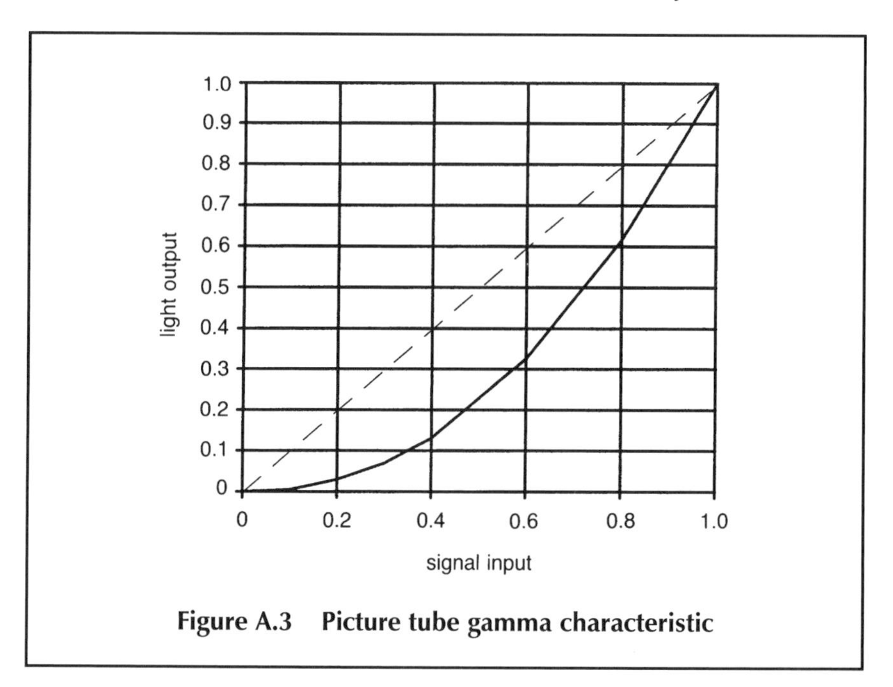

Figure A.3 Picture tube gamma characteristic

tion was made as early as possible. Cameras have therefore gamma correction added to compensate for the gamma of the picture tube.

By pre-distorting gamma at source to correct an error at the destination means the signal chain is 'gamma corrected'. All picture sources must, therefore, have a gamma of 0.4545 and all picture displays must have a gamma of 2.2 (whether CRT-based or not). The overall relationship of light input to camera, to light output of the screen must be linear, making overall gamma unity.

Gamma correction curve is shown in Figure A.4. In practice, the correction is modified for circuitry design reasons. True correction requires a rise from 0% that is vertical, requiring a gamma processor gain of infinity, an impossibility. So the initial gain is limited to 5 up to a signal output of 20%. From 20% the curve follows a gamma of 0.4. This approximation is perfectly adequate and has been adopted as a standard. Camera manufacturers have designed gamma correction in slightly different ways. Some industrial cameras use an initial rise of 3½:1, giving a slight increase in dark tone density rather similar to raising gamma to 0.45 in a broadcast camera.

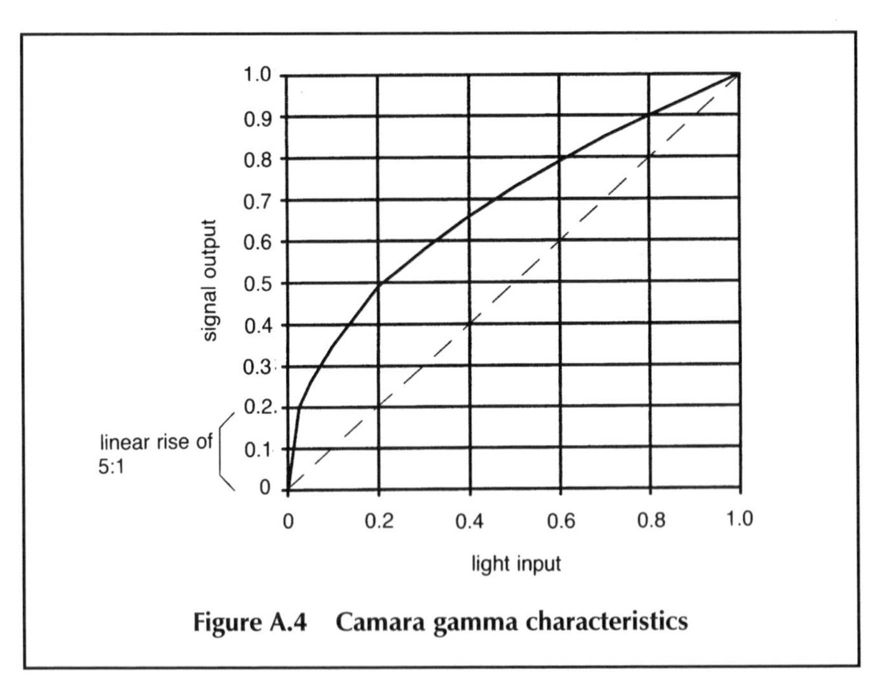

Figure A.4 Camara gamma characteristics

Although a gamma of 0.4 is now a broadcast standard, lighting directors and photographers, seeking greater creative control, alter camera gamma to suit their own specific purposes. For example, raising gamma gives progressively greater dark tone density but at the expense of lighter tone gradation.

3. Contrast range

This defines how the scene tonal range of the system, usually as applied to cameras. The input/output relationship of a typical camera is set out in Table A.1 and shows the nonlinearity of gamma correction.

Table A.1

Light input	*Signal output*	
100%	(90% reflectance white card)	100% (clip level)
50%	–1 stop	75%
25%	–2 stops	55%
18%	(18% reflectance grey card)	50%
12.5%	–3 stops	40%
6.25%	–4 stops	25%
3.125%	–5 stops	3% (pedestal level)

The contrast range of the camera is therefore five stops, where one stop is equivalent to doubling or halving light input. The camera lens is calibrated in stops. At black (or pedestal) the signal becomes lost in system noise; as noise figures improve so effective contrast range increases. But, it must be remembered, that to appreciate any extension of the contrast range, the picture display and viewing conditions must be upgraded accordingly.

Extending contrast range by compressing higher tones, as with the 'knee' characteristic, is less dependent on viewing conditions. Using lower gamma has similar, but not identical, effect whilst a higher gamma does the reverse.

Raising pedestal also compresses the signal. For instance, by raising to 10% the useful signal range will occupy only 90% of the available system contrast range. The scene contrast remains as before.

4. Gain and signal levels

Gain, or amplification, expresses the increase or decrease in signal level, or amplitude. Gain is specified in decibels (dB's) or as a factor. It may also appear, when related to light levels, as light stops.

Decibels, dB's	–6	0	6	9	12	18
Gain (factor)	×½	×1	×2	×3	×4	×8
Sensitivity (f-stops)	–1	0	+1	+1½	+2	+3

Decibels are a logarithmic function with origins in audio measurement. Its use in video has come about because of its technical convenience and is often used when quoting video signal levels.

An alternative to quoting signal level in percentage is to use IRE units. The figures are the same.

Percentage, or IRE units	0	2	5	10	20	50	70	85	100
Volts (ref to video black)	0	0.014	0.035	0.07	0.14	0.35	0.49	0.595	0.7
Volts (ref to sync level)	0.3	0.314	0.335	0.37	0.44	0.65	0.79	0.895	1.0

IRE units are also used to state levels below black, sync amplitude is an example.

5. Standard signal values

NTSC and PAL have been taken, for the purpose of this book, as having the same signal values. The actual values as set out in Figure A.5 and Figure A.6.

PAL is by convention, measured in volts. NTSC is in IRE values which are based on percentage of one volt. Note that the NTSC and PAL amplitudes are slightly different: 100 IRE is 714 mV (0.714 V) and 40 IRE is 286 mV (0.286 V), giving a slightly larger picture-to-sync ratio. Reference black is at 7.5 IRE, this is the standard NTSC pedestal.

Field sync is a sequence of half line frequency pulses known as the broad pulses because of their length. Equalising pulses are the half line pulses either side of the field sync that ensure the field sync is in the same relative position every field.

Some of the unused blank lines after the field sync may be used for test signals, video tape timecode, identification and information.

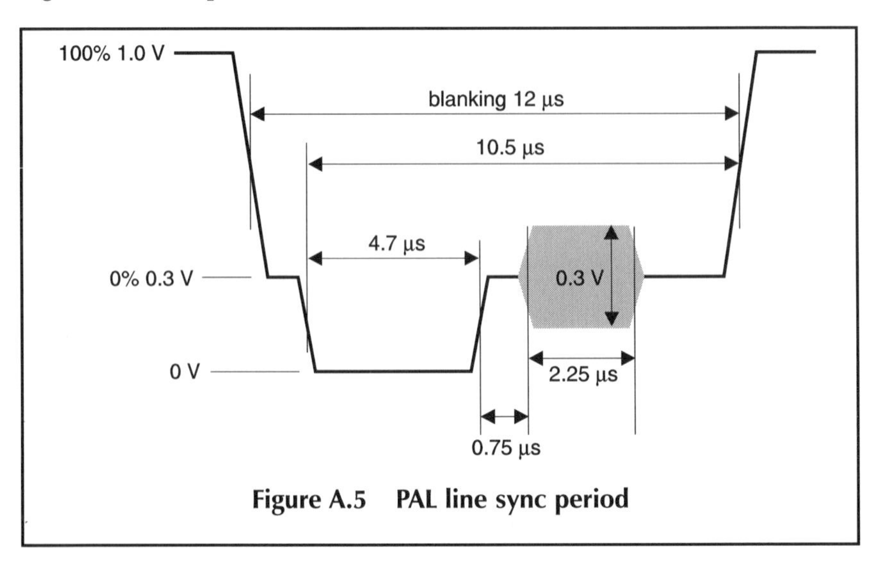

Figure A.5 PAL line sync period

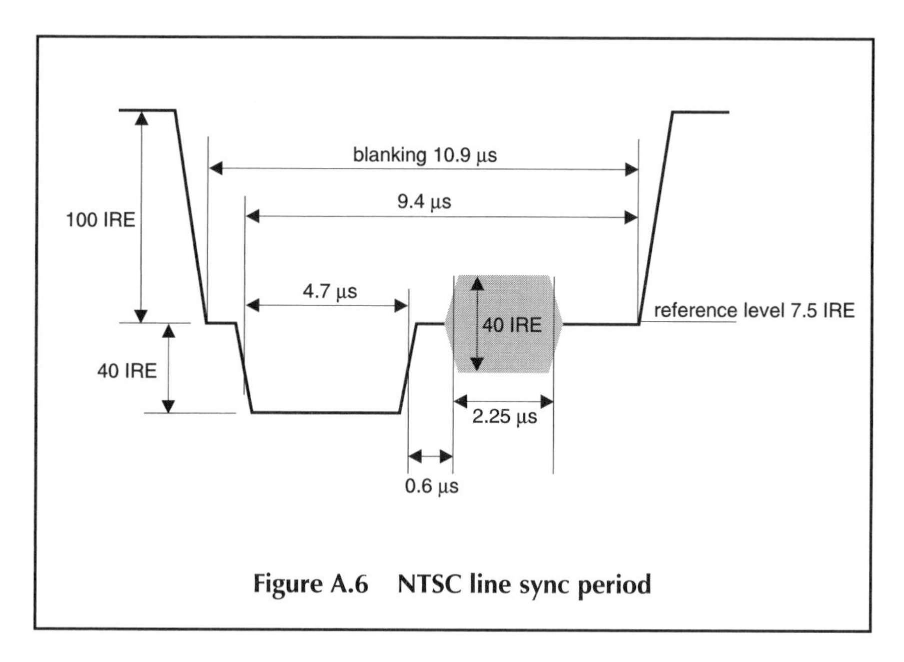

Figure A.6 NTSC line sync period

6. Light measurement

Light power is measured in candelas, or candle power. The light leaving the candle to illuminate a surface area is measured in lumens. The illumination of one candle power at one foot is one foot-candle, or one lumen/sq. foot and is the standard Imperial unit of illumination. The metric equivalent is the lumen/m^2, or lux.

Therefore:

1 foot candle = 1 lumen/ft^2 = 10.76 lux

Light meters (or exposure meters) are usually calibrated in foot candles or lux, sometimes both.

Typical studio light levels are 80 foot-candles, 860 lux. Modern cameras can work down to 10 lux and lower, but such light levels are not practical for normal studio operations.

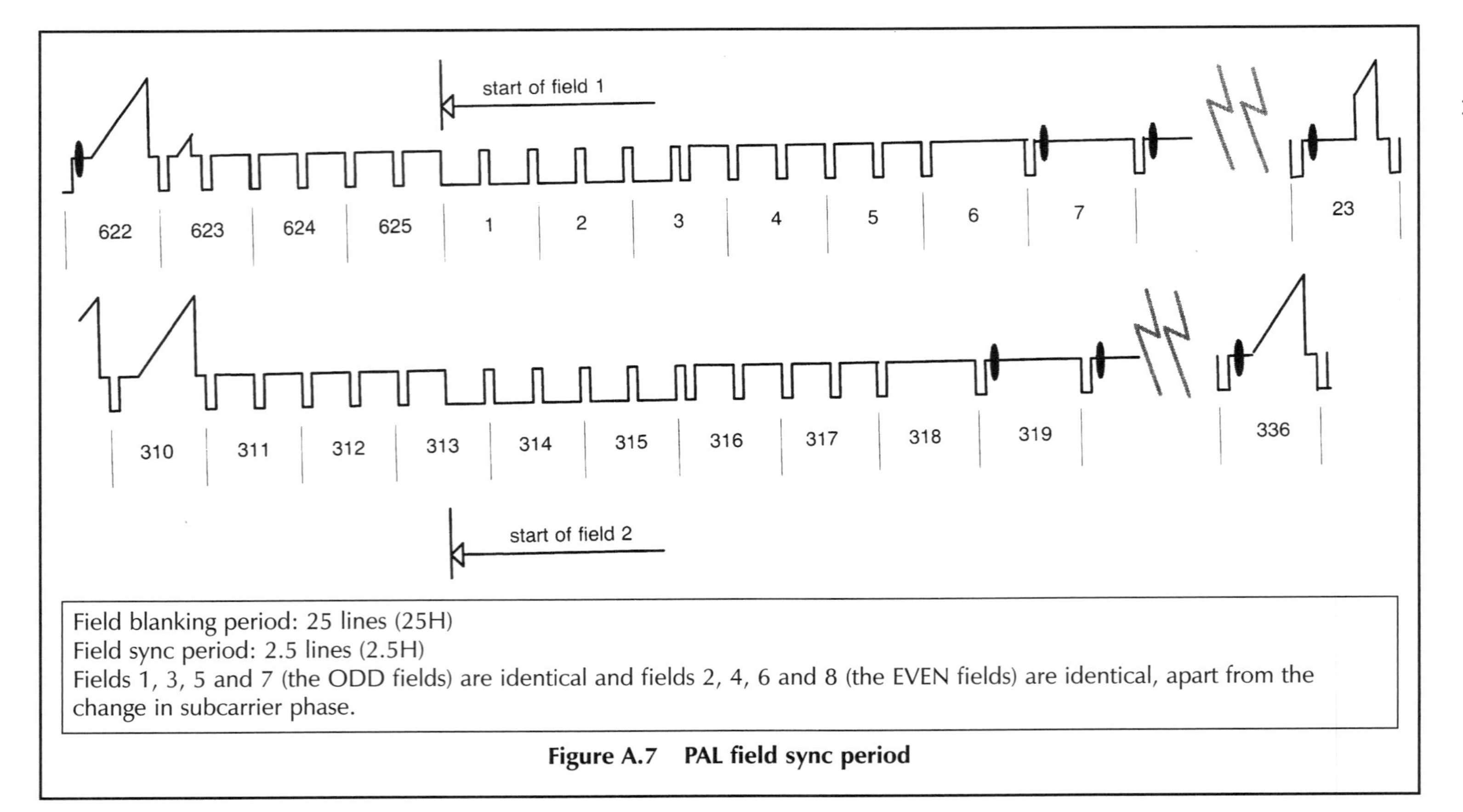

Figure A.7 PAL field sync period

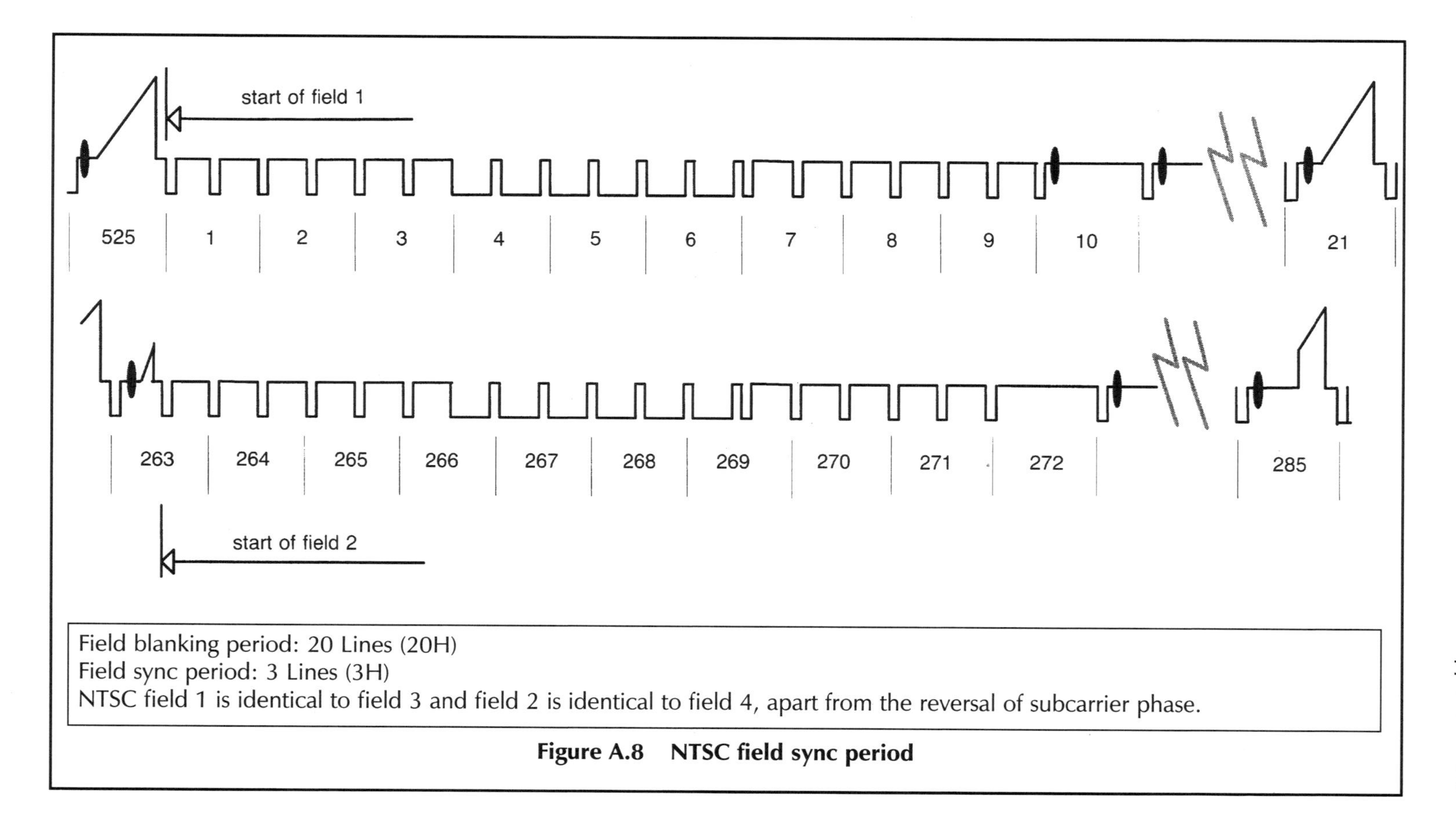

Figure A.8 NTSC field sync period

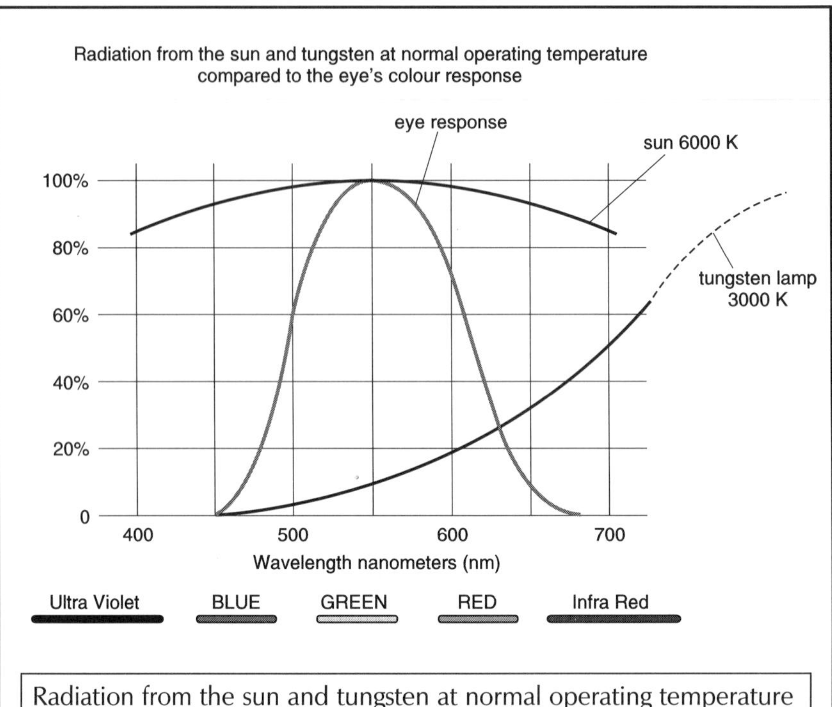

Radiation from the sun and tungsten at normal operating temperature compared to the eye's colour response.

Figure A.9 Relative spectral response

7. Colour

Colour temperature

This is the relationship of light colour to temperature. Any body, on being raised in temperature, emits radiation. At first this is only heat but as the temperature rises so the body starts to glow. Heat is infra red radiation, which is invisible to the human eye. As temperature rises the body glows, first red, then orange, then yellow. It is seen as white when it approaches the sun's temperature at 6000K.

Everything on earth radiates heat; it is only at absolute zero, that is zero Kelvin (–273.15°C), that radiation ceases.

At 3000K tungsten glows orange/yellow. The eye continually balances colour to normalise, similar to the camera white balancing, and is generally unaware of the colour differences between tungsten lighting and daylight (Figure A.9). Tungsten lighting is flexible because it is easily varied in brightness. It does, at the same time, alter colour for the reasons set out above. Operating a tungsten lamp about 80% output gives useful adjustment either side, giving a typical colour temperature of 2900K. and much longer lamp life than when worked at full output.

Colour temperature may be adjusted by light filtration, either at the camera, or the lamp. Correction filters from tungsten to daylight, 3000K to 6000K. Daylight to tungsten are also available, as are intermediate values for fine adjustments to match lamp colours.

Colorimetry

The science of colour. The principle of additive colour as used in television is based on numerical colour values applied to the primaries, red, green and blue. The unit of light is the lumen and any colour may be defined as:

$n(C) \equiv x(R) + y(G) + z(B)$ where n, x, y and z are in lumens

The standard CRT phosphors used in television deliver 0.3 lumens of red, 0.59 lumens of green and 0.11 lumens of blue, to produce one lumen of white. These are normalised to equal values such that:

1 unit of R + 1 unit of G + 1 unit of B $\equiv$ 1 unit of white

Hence, the equal values RGB used by the video standard. The conversion factors from lumens to equal values is introduced in the camera, the reverse taking place at the screen.

Once equal values are in place, it is easy to define any colour and its luminosity by matching it with specific values of the three primaries.

A colour may be seen as specified by the co-ordinates X and Y. These are derived from the CEI colour diagram. Further reading is recommended for a more complete understanding of the subject.

8. Termination

A video circuit consists of a source driving a destination via a cable. The source has a resistance of 75 ohm and the destination terminates with a load of 75 ohm. These are specified as resistors. The cable has a characteristic impedance of 75 ohm, impedance is the combination of resistance and reactive (capacitive and/or inductive) elements. Figure A.10 refers.

Current driven by the source voltage travels around the circuit through both 75 ohm resistors. Each resistor dissipates half the circuit power, therefore half is lost at the source and half received at the destination. But because the cable source and termination requirements are met, there is minimal cable loss and distortion.

9. DC restoration

It is not possible in a practical circuit to maintain a DC circuit. The input circuit in Figure A.11 has a capacitor C to protect the processing circuitry from excessive DC variations arising from circuit interference and noise. Resistor R assists with this action. The capacitor does not pass DC and DC component of the video signal is therefore removed.

Replacing R with a switch that operates in back porch restores the DC by bringing the capacitor to black level at the start of every line. The switch closes for the duration of the clamp pulse and is designed to avoid distorting the colour burst, Figure A.12.

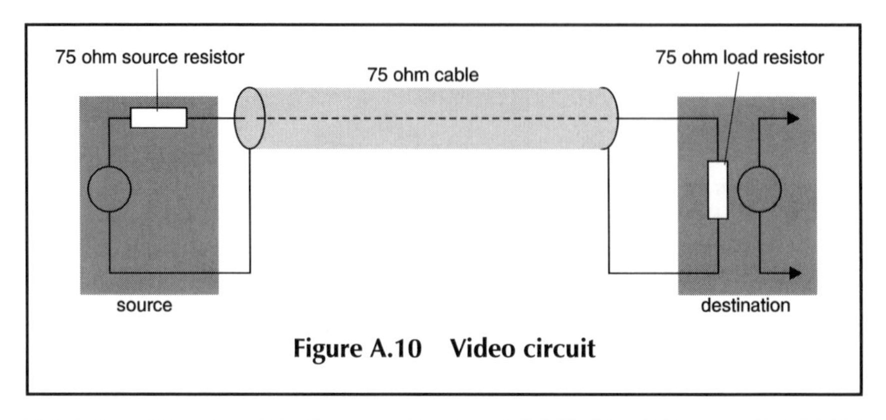

Figure A.10 Video circuit

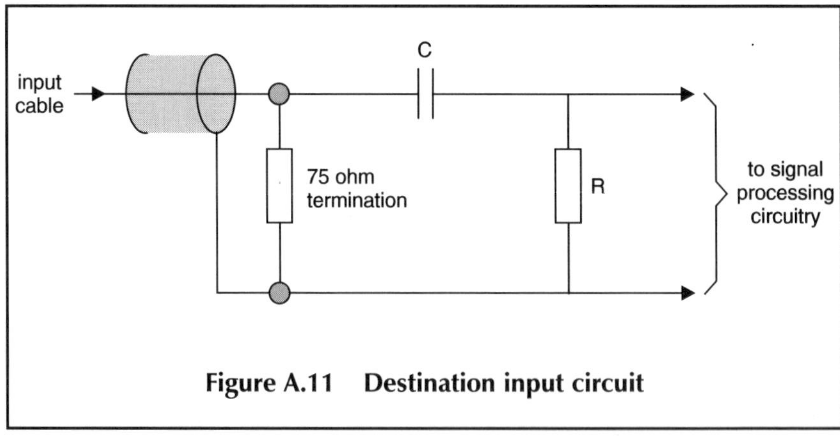

Figure A.11 Destination input circuit

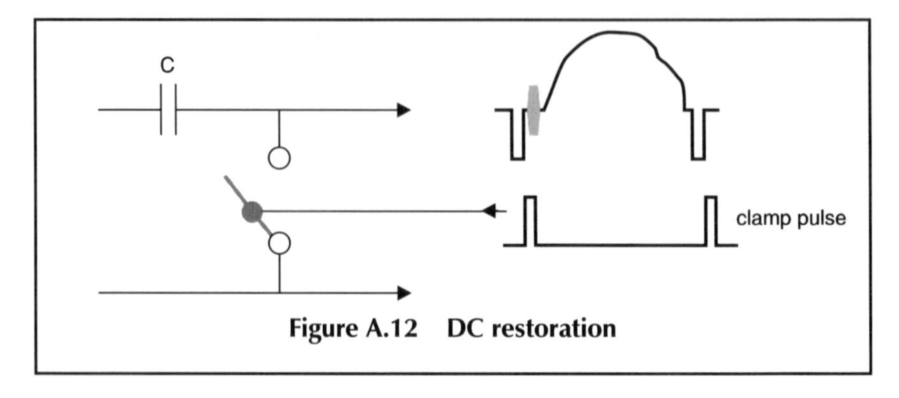

Figure A.12 DC restoration

GLOSSARY

Italics refer to *index* or other *glossary* entries.

AC *Alternating* current. Electric current that changes direction of flow. See *sine wave* and *DC*.

Amp The standard unit of electric *current*. See *voltage*.

Amplifier A device to increase signal *level*. To do so requires the supply of *power*. See *Proc Amp*.

Amplitude Amount of *signal*. See *level* and Index.

Bandwidth See Index.

Battery A source of electrical energy held in chemical form. Also used to denote a battery of electric cells. See *power supply*.

Binary See Index.

BIT *Binary* digit. 0 or 1.

BNC The standard video bayonet locking coaxial *connector* .

Breezeway Portion of the video signal between trailing edge of line sync and start of colour burst. See Index *back porch*.

Broad pulses The field sync pulses. See Index.

Byte Digital number or word. A bit sequence.

Cable Wire-based circuit.

Capacitor An electric component. Holder of an electric *charge*. A capacitor has an *insulator* that stops the passage of *DC*. *AC* flow increases as its frequency rises. See *inductor*.

CCD Charge coupled device. A silicon *pixel* array onto which the image is focused so building up an equivalent electric *charge* that is read off to form the video signal.

CCS Colour subcarrier. See Index.

CEI diagram The colour co-ordinates chart that defines colour.

Charge An electric quantity that is not flowing.

Circuit What an electric *current* flows in. See *transmission circuit* and *circuitry* and Index.

Circuitry The electronic circuits inside equipment. See *circuit*.

Colour black *Black* and *burst*. See Index.

Coaxial Two-wire cable where one conductor is totally enclosed by the other, known as inner and outer (or screen). Also coaxial connector, see *BNC*.

Compression Reducing signal *bandwidth* or *memory* requirements by sending only essential information, e.g. only sending picture movement information.

Configure To construct, or set up, a system to perform a particular task. See *jackfield*.

Conforming Where *component* source play-in material is edited down to *composite* for transmission.

Connector The means of connecting cable to apparatus, etc.

Current The movement of electricity. Current only exists when a voltage is present to drive it around a circuit.

DC Direct current. *Current* flow that is constant and does not change direction. See *AC*.

Decibels Logarithmic unit of voltage, current or power difference, or change.

dB = 20 [log (voltage OR current difference)] =
10 [log (power difference)]

dB Abbreviation of *decibel*.

EBU European Broadcasting Union.

Equalising pulse The pulses occurring either side of the field sync.

Genlock Generator lock. The method of video synchronising by slaving one piece of equipment to another. See Index.

Gain Increase (or decrease) of signal *level* expressed as a factor, or in *decibels*, that is achieved by a *circuit*, *amplifier*, or other piece of equipment.

Gigabyte One thousand million *bytes*.

Horizontal sync Another name for *line sync*.

Inductor The opposite of *capacitor*. The inductor passes *direct current* but opposes changes of current flow. *AC* flow increases as its frequency falls. See *capacitor*.

Impedence The control of an electric current by *resistance* and *reactance* separately or combined. The term is used whenever a *circuit* is not pure *resistance*.

Insulator Stops the flow of electric *current*. May be described as infinite *resistance*.

IRE Institute of Radio Engineers. 'Percentage Unit' of signal level, or 1/140th of 1 V.

ITS Insertion test signal. A test signal inserted into the field blanking period.

Jackfield The means of interconnecting sources and destinations, apparatus and circuits. Enables system interchange or re-configuration. See *Configure*.

Kilobyte One thousand bytes. See *byte*.

LCD Liquid Crystal Display. A compact and low power image display that works by changing its light transmission with applied voltage.

Level Amount of signal, normally stated as *voltage*, occasionally as *power*. See *amplitude* and Index.

Line Alternative term to *circuit* but more usually denoting a *cable* based circuit.

Line rate The *frequency* of the video picture line structure.

Memory The electronic storage of information.

Microprocessor Digital calculation or manipulation chip.

Microsecond One millionth of a second.

Millisecond One thousandth of a second.

Monochrome See Index.

MPEG Motion Picture Expert Group. A US advisory body to the industry.

Musa Standard *coaxial* slide-fit *connector* as used on *jackfields*, longer than the BNC but smaller diameter.

Nanosecond One thousand millionth of a second.

NTSC National Television System Committee.

Ohm See Index.

Oscillator A generator of *alternating* signals, usually *sine wave*, but not necessarily so.

Oscilloscope See Index.

PAL Phase Alternate Line.

Phase Measure of *synchronism*, or signal coincidence. Relative timing of *signals* or *sine waves*.

Power The amount of energy transferred in a given time. Product of *voltage* and *current* and measured in watts. See also *decibels*.

Power supply Source of electric power as supplied to equipment. See *Battery*.

Proc amp Processing amplifier. In video systems it restores *synchronising pulses* and *colour burst* to their correct shape and position. Provides *black level*, *gain* and *peak white clipping* adjustment. Usually has a *gain* of unity, i.e. does not amplify. See *amplifier*.

Raster The *scanning* structure of a video picture.

Reactance The current controlling effect provided by a *capacitor* or *inductor*.

Resistance The current controlling effect offered by a *resistor*, cable or wire. Measured in *ohms* its effect applies equally to *AC* and *DC*. See *ohm*.

Resistor An electric component that is used to provide *resistance* to *current* flow.

Rise time See Index.

Ramp signal See Index.

Signal See Index.

Sine wave See Index.

Time Time, according to the special theory of relativity, is not absolute; its rate can vary from observer to observer. However, for the purposes of this book, time is taken as absolute.

Transfer characteristic How the signal transfers from one medium to another affects a signal.

Transmission Sending a signal down a *line* or *circuit*. Transmission also describes sending by *wireless*. Also broadcasting a TV or radio programme.

Transmission circuit The means by which a signal is transmitted. Longer ones may include amplifiers and other apparatus.

Vertical interval The period between fields.

Vertical sync Another name for *field sync*.

Volt Unit of *voltage*.

Voltage Electric force, e.g. from a *battery* or *power supply*, or a *signal* source. The potential to drive a *current* in a *circuit*.

Weighting Where a signal is pre-distorted or changed to make it more representative of what the eye sees. Or, to make the effects of circuit conditions less obvious.

Wireless Sending without wires. Radio *transmission*.

Signal abbreviations

The following are in common usage, but alternative versions sometimes appear.

B and B Black and burst (sync and colour burst).

BS Black and sync (sync only).

CBC Colour, black and sync (sync and colour burst).

CVBS Colour, video, black and sync (complete composite video).

CYMK Cyan, yellow, magenta and black (standard computer/printer interface).

RGB Red, green and blue (as separate circuits).

TTL Transistor/transistor logic (the standard digital circuit interface).

YC Luminance and chrominance (as separate circuits).

BIBLIOGRAPHY

Craig, M. (1991 & 1994) *Television Measurements*, Tektronix Inc.

D'Amato, P. (1984) *Study of the Effects of Various Impairments on the 20T Pulse*, European Broadcasting Union.

Gressman, R. (1978) *Guiding Principles for the Design of Television Waveform Monitors*, European Broadcasting Union.

Henderson, H. (1964) *Colorimetry*, Engineering Training Department, BBC.

Hodges, P. (1993) *Vical. a Video Calibration System Using a Standard Picture Monitor*, The Journal of Photographic Science.

Hodges, P. (1994) *The Video Camera Operator's Handbook*, Focal Press.

Sims, H. V. (1968) *The Principles of PAL Colour Television*, Engineering Training Department, BBC/Iliffe.

Watkinson, J. (1994) *An Introduction to Digital Video*, Focal Press.

INDEX